Ralph J. Fessenden
University of Montana

and Joan S. Fessenden

ORGANIC CHEMISTRY

STUDY GUIDE WITH SOLUTIONS

Willard Grant Press
Boston, Massachusetts

Designed by David Chelton. Text composition by Rita de Clercq Zubli on an
IBM Executive Typewriter. Additional technical art by Joan Goodman.
Cover design by John Servideo. Printed and bound by Hamilton Printing Co.

To the Student

This Study Guide is intended to accompany the text <u>Organic Chemistry</u>. Included in this guide are some hints to help you study organic chemistry; a chapter-by-chapter discussion of many important features; and answers to the study problems at the end of each chapter in the text. (The answers to the study problems <u>within</u> the chapters are at the end of the text.) In this guide, we have included explanations of the answers where appropriate. These explanations will be especially helpful if you have difficulty answering a particular type of problem.

Sometimes organic nomenclature, bonding, reactions, etc. seem perfectly clear to a student, yet he does poorly on an examination. To avoid being in this situation, use your study time correctly.

What is the correct way to study organic chemistry? It is similar to learning a foreign language: first, understand the material; then, memorize the salient features. Rote memorizing of organic chemistry without first understanding the material is a waste of time. On the other hand, studying only to the point of understanding does not result in good examination grades. Only by coupling understanding and memorizing can you progress to thinking in the language of organic chemistry.

As a check on your understanding, you should work the problems within the chapter as you go along. (Cover the answers to Sample Problems with a sheet of paper, and try to work them as well.)

It is important to <u>write</u> your answers to the problems rather than simply to envision the answers in your mind. There are two reasons for writing out the answers: (1) for practice in the mechanical art of writing organic formulas, and (2) for reinforcing the subject material.

After you have studied the textual material and your lecture notes thoroughly, tackle the chapter-end problems. Some of these problems are drill, while some require thought and ingenuity. See if you can answer the problems yourself, and <u>then</u> check the answers and the text. (If you try to take shortcuts, you are only deceiving yourself.) If your answers are correct, you have a reasonable mastery of the material. If you have several incorrect answers, go back and restudy the chapter, your lecture notes, and the hints in this guide.

As a starting point on what to study, be sure you understand each definition and each reaction in the chapter summary. Be sure you can extrapolate from the generalities of the summary to the specific reactions mentioned in the chapter. Then, be sure you can extrapolate from the specific reactions in the chapter to reactions of new and different compounds.

General: $RCO_2H + OH^- \rightleftharpoons RCO_2^- + H_2O$

Specific: $CH_3CO_2H + OH^- \rightleftharpoons CH_3CO_2^- + H_2O$

Extrapolated:
$$CH_3\overset{\underset{|}{CH_3}}{C}H\overset{\underset{|}{CH_3}}{C}HCHCO_2H + OH^- \rightleftharpoons CH_3\overset{\underset{|}{CH_3}}{C}H\overset{\underset{|}{CH_3}}{C}HCHCO_2^- + H_2O$$

Organic molecules are three-dimensional. Besides a pencil and paper, you will need a molecular model kit. At the start of the course, especially in discussions of stereochemistry, you will find it helpful to construct models of organic molecules and compare them with their representations on paper.

Contents

Chapter 1

Atoms and Molecules— A Review

Some Important Features

Electrons are found in shells surrounding the nucleus of an atom. Each shell is composed of one or more atomic orbitals. The first shell contains a 1s orbital (spherical); the second shell contains one 2s orbital (spherical) and three 2p orbitals (dumbbell-shaped and mutually perpendicular). Each orbital can hold zero, one, or two electrons. Electrons are usually contained in the lowest-energy orbitals possible (1s, then 2s, then 2p).

The halogens, oxygen, and nitrogen have fairly high electronegativities (attraction for outer electrons). The metals have low electronegativities, while carbon and hydrogen have intermediate electronegativities.

Chemical bonds are formed by electrons in the outer shell of an atom. Whether ionic or covalent bonds are formed depends on the electronegativity difference between two atoms. Carbon forms covalent bonds with other elements. These covalent bonds may be nonpolar ($C-C$ or $C-H$) or polar ($C-O$, $C-N$, or $C-Cl$), depending on the electronegativity difference between C and the other element.

Molecules with NH, OH, or HF bonds can form hydrogen bonds with each other or with other molecules containing N, O, or F atoms with unshared electrons.

An acid is a compound that can donate H^+ or accept electrons, while a base is a compound that has unshared electrons that can be donated. Common organic acids are the carboxylic acids, compounds with a $-CO_2H$ group. Common organic bases are amines, compounds with a nitrogen atom bonded to three other atoms.

$$RCO_2H \quad + \quad R_3N: \quad \rightleftharpoons \quad RCO_2^- + R_3\overset{+}{N}-H$$

<u>a carboxylic acid</u> <u>an amine</u>

The strengths of acids and bases are indicated by their pK_a values or pK_b values. A smaller numerical value for pK_a means a stronger acid.

Other important topics covered in this chapter are ionic and covalent bonds, calculation of formal charge, types of chemical formula, bond distances and angles, and bond dissociation energies.

Reminders

In stable compounds, carbon forms four bonds, hydrogen forms one bond, and oxygen forms two bonds.

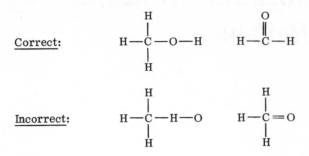

Hydrogen bonds are formed between unshared electrons on O, N, or F and a hydrogen attached to O, N, or F.

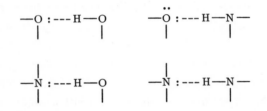

Answers to Problems

1.15   (a) $1\underline{s}^2\ 2\underline{s}^2\ 2\underline{p}^2$ (b) $1\underline{s}^2\ 2\underline{s}^2\ 2\underline{p}^6\ 3\underline{s}^2\ 3\underline{p}^2$

(c) $1\underline{s}^2\ 2\underline{s}^2\ 2\underline{p}^6\ 3\underline{s}^2\ 3\underline{p}^3$ (d) $1\underline{s}^2\ 2\underline{s}^2\ 2\underline{p}^6\ 3\underline{s}^2\ 3\underline{p}^4$

1.16 $2\underline{s}$ and $2\underline{p}$

1.17 $90°$, which is the angle between $\underline{p}$ orbitals.

1.18   (a) Si   (b) B   (c) C

1.19   (a) O   (b) N   (c) C   (d) O

1.20   (a) $CH_3CO_2^-$ Na^+ (b) all covalent   (c) Li^+ $\ ^-OH$

           covalent ionic   ionic covalent

(d) CH_3O^- Na^+ (e) all covalent   (f) Br^- Mg^{2+} $\ ^-OH$

           covalent ionic   ionic   covalent

(g) covalent   (h) covalent

1.21   Li, 1; Be, 2; B, 3; C, 4; N, 3; O, 2; F, 1; Ne, 0

1.22   four. Each covalent bond is two electrons; a second-period element can accommodate a maximum of eight electrons.

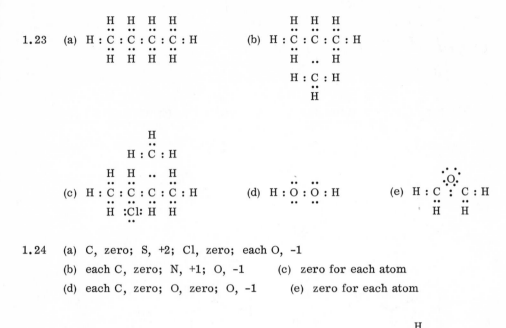

1.23

(a)
$$\overset{\displaystyle H \ \ H \ \ H \ \ H}{H : C : C : C : C : H}$$
$$\underset{\displaystyle H \ \ H \ \ H \ \ H}{}$$

(b)
$$H : C : C : C : H$$

(c)
$$H : C : C : C : C : H$$

(d)
$$H : \ddot{O} : \ddot{O} : H$$

(e)
$$H : C : C : H$$

1.24 (a) C, zero; S, +2; Cl, zero; each O, -1

(b) each C, zero; N, +1; O, -1   (c) zero for each atom

(d) each C, zero; O, zero; O, -1   (e) zero for each atom

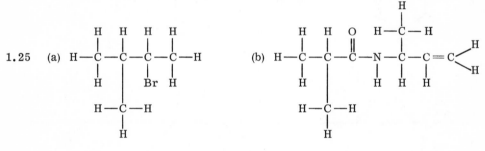

1.25 (a)

(b)

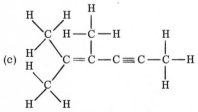

(c)

Unless they are specifically called for, do not show unshared electrons in structural formulas.

1.26 (a) CH_3OH (b) $CH_3CH_2NH_2$ (c) $(CH_3)_2NH$

1.27 All are $C_4H_{10}O$.

1.28 (a)

(b)

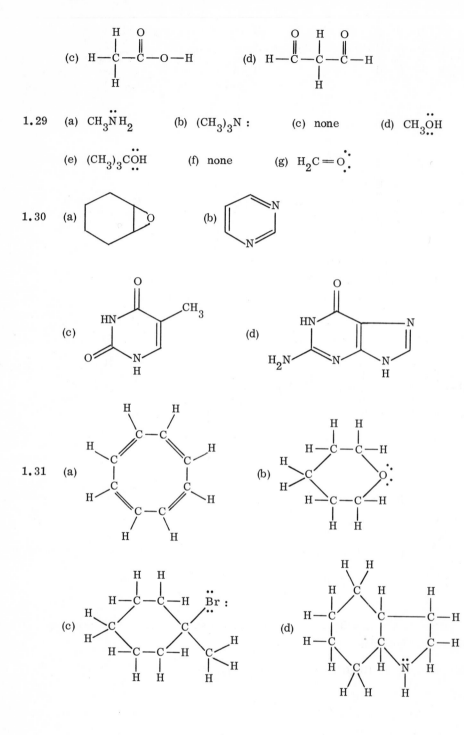

(c) H—C—C—O—H   (d) H—C—C—C—H

1.29   (a) $CH_3\overset{..}{N}H_2$ (b) $(CH_3)_3N:$ (c) none   (d) $CH_3\overset{..}{\underset{..}{O}}H$

       (e) $(CH_3)_3C\overset{..}{\underset{..}{O}}H$ (f) none   (g) $H_2C=\overset{..}{\underset{..}{O}}:$

1.30   (a)   (b)

       (c)   (d)

1.31   (a)   (b)

       (c)   (d)

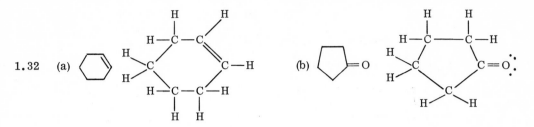

1.32   (a)

(b)

A formula with the double bond or carbonyl group in a different position represents the same compound and is also correct. For example:

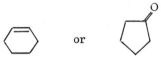

or

$$\underline{\underline{\Delta H}}$$

1.33   (a) F—F   ⟶   2 F·   +37

   H_3C—H   ⟶ CH_3· + H·   +104

   CH_3· + F·   ⟶ CH_3F -108

   H· + F·   ⟶   HF $\underline{-135}$
                                    -102 kcal/mole

$$\underline{\underline{\Delta H}}$$

   (b) Cl—Cl   ⟶   2 Cl·   +58

   CH_3—H   ⟶ CH_3· + H·   +104

   CH_3· + Cl· ⟶ CH_3Cl -83.5

   H· + Cl·   ⟶   HCl $\underline{-103}$
                                    -24.5 kcal/mole

$$\underline{\underline{\Delta H}}$$

   (c) Br—Br   ⟶   2 Br·   +46

   CH_3—H   ⟶ CH_3· + H·   +104

   CH_3· + Br· ⟶ CH_3Br -70

   H· + Br·   ⟶   HBr $\underline{-87}$
                                    -7 kcal/mole

$$\underline{\underline{\Delta H}}$$

(d) $I—I \longrightarrow 2\ I\cdot$ +36

   $CH_3—H \longrightarrow CH_3\cdot + H\cdot$ +104

   $CH_3\cdot + I\cdot \longrightarrow CH_3I$ -56

   $H\cdot + I\cdot \longrightarrow HI$ $\underline{-71}$

                                   +13 kcal/mole

1.34   (a) $CH_3CH_2—Cl \longrightarrow CH_3CH_2\cdot + Cl\cdot$ or $CH_3CH_2^+ + Cl^-$

(b) $H—OH \longrightarrow H\cdot + \cdot OH$ or $H^+ + OH^-$

(c) $H—NH_2 \longrightarrow H\cdot + \cdot NH_2$ or $H^+ + {}^-NH_2$

(d) $CH_3—OH \longrightarrow CH_3\cdot + \cdot OH$ or $CH_3^+ + OH^-$

(e) $CH_3O—H \longrightarrow CH_3O\cdot + \cdot H$ or $CH_3O^- + H^+$

In heterolytic cleavage, the bonding electrons always go with the more electronegative atom.

1.35   (a) $\overset{\delta+}{CH_3}—\overset{\delta-}{\underset{}{\textcircled{O}}}—\overset{\delta+}{H}$ (b) $CH_3\overset{\overset{\delta-}{\overset{\displaystyle O}{\|}}}{\underset{\delta+}{C}}CH_3$ (c) $\overset{\delta-}{\textcircled{F}}—\overset{\delta+}{CH_2}CO_2H$

(d) $(CH_3)_2N\overset{\delta+}{CH_2}CH_2—\overset{\delta-}{\textcircled{O}}—\overset{\delta+}{H}$

In (a) and (d), the most electronegative atom is bonded to two other atoms; therefore, it is necessary to show the polarity of both of its bonds.

1.36   (a) $\overset{\delta-}{C}—\overset{\delta+}{Mg}$ (b) $\overset{\delta+}{C}—\overset{\delta-}{Br}$ (c) $\overset{\delta+}{C}—\overset{\delta-}{O}$ (d) $\overset{\delta+}{C}—\overset{\delta-}{Cl}$

(e) $\overset{\delta-}{C}—\overset{\delta+}{H}$ (f) $\overset{\delta-}{C}—\overset{\delta+}{B}$

The direction of the dipole is determined by comparison of the electronegativities of the bonded atoms.

1.37   (a) $CH_3CH_2CH_3$, $CH_3CH_2CH_2NH_2$, $CH_3CH_2CH_2OH$

(b) $CH_3CH_2CH_2I$, $CH_3CH_2CH_2Br$, $CH_3CH_2CH_2Cl$

The ranking is based upon differences in electronegativities of atoms in the polar groups. A greater electronegativity difference means a more polar molecule.

1.38   (a) < (b) < (c). In problems of this type, it is often helpful first to pick out the answer at one end of the scale, then the one at the other end. The intermediate cases then can be handled more easily. In Problem 1.38, the compound with least ionic character is CH_4, which contains nonpolar bonds. The compound with the most ionic character is LiCl, which is an ionic compound. The intermediate case is HBr, which contains a polar covalent bond.

1.39   (a), (b), and (c), but not (d), which has unshared electrons, but does not have a hydrogen atom bonded to N, O, or F.

1.40   (a) $CH_3CH_2O\overset{\text{H}}{\text{H}}$---$OCH_2CH_3$ (b) none   (c) none   (d) none

1.41   with itself: (a), (e);   with water: (a), (b), (d), (e)

Any compound that can hydrogen bond with itself can also hydrogen bond with water. Compounds (b) and (d) have unshared electrons that can form a hydrogen bond with the partially positive hydrogen atom of water.

1.42   (a) $(CH_3)_2N\overset{\text{H}}{}$---$N(CH_3)_2$ (b) $(CH_3)_2NH$--- OH_2

    (c) $(CH_3)_2N\overset{\text{H}}{}$---$H_2O$ (d) H_2O ---H_2O

Nitrogen is less electronegative than oxygen; therefore, its unshared electrons are more loosely held and more available for hydrogen bonding. The OH bond is very polar (and the H is more positive) because of the high electronegativity of oxygen. For these two reasons, the N---HO hydrogen bond in (c) is the strongest hydrogen bond in the solution.

1.43   (a) $CH_3OH + {}^-OH$ (b) $CH_3NH_3^+ + Cl^-$ (c) ⬡—$CO_2^- + H_2O$

    (d) ⬡NH_2^+ (e) CH_3CO_2H (f) $CH_3NH_3^+ + CH_3CO_2^-$

1.44   (a) 4.75   (b) 9.4   (c) 4.3   (d) ~16   (e) ~43

    (e) < (d) < (b) < (a) < (c)

The calculation of pK_a values is discussed in Section 1.10A of the text. These values may also be obtained directly from a calculator with a log function. (Remember, however, to change the sign: $pK_a = -\log K_a$)

1.45   (a) 13   (b) 9.4   (c) 3.2   (d) 5.8   (a) < (b) < (d) < (c)

1.46   (a) Na   (b) Cl   (c) Ne

1.47   (b) and (c), because they have the same numbers of electrons in their outer shells.

1.48   Either a mixture of NaBr and LiCl or a mixture of NaCl and LiBr in water yields Na^+, Br^-, Li^+, and Cl^-. A solution of CH_3Cl and NaBr yields CH_3Cl, Na^+, and Br^-. This is different from a solution of CH_3Br and NaCl, which yields CH_3Br, Na^+, and Cl^-.

1.49 $PbCl_4$ is a covalent compound, while $PbCl_2$ is ionic.

1.50   (a) ⁻CN, because carbon has a formal charge of –1 and N has a formal charge of 0.

(b) ⁻C≡CH, because the left-hand carbon has a formal charge of –1 and the right-hand carbon has a formal charge of 0.

(c) $\overset{+}{C}H_2OH$, because carbon has a formal charge of +1 while all other atoms have formal charges of 0.

(d) CH_3O^-, because oxygen has a formal charge of –1 and the other atoms have formal charges of 0.

1.51   (a) △   (b) (with O on top)   (c) (five-membered ring with O)

Four- and three-membered rings would also be correct for (c). For example,

(square with OH)   or CH_3—(triangle)—OH

$$\underline{\underline{\Delta H}}$$

1.52   (a) Cl_2 ⟶ 2 Cl·   +58

CH_3CH_2—H ⟶ CH_3CH_2· + H·   +98

CH_3CH_2· + Cl· ⟶ CH_3CH_2Cl –81.5

H· + Cl· ⟶ HCl $\underline{–103}$

–28.5 kcal/mole

$$\underline{\underline{\Delta H}}$$

(b) Br_2 ⟶ 2 Br·   +46

CH_3CH_2—H ⟶ CH_3CH_2· + H·   +98

CH_3CH_2· + Br· ⟶ CH_3CH_2Br –68

H· + Br· ⟶ HBr $\underline{–87}$

–11 kcal/mole

The reaction in (a) liberates more energy.

1.53   (a) $CH_3O^- + H^+$ (b) $(CH_3)_2\overset{+}{C}H + Br^-$ (c) $CH_3\overset{-}{C}H_2 + Li^+$

1.54   a planar molecule: (BF₃ structure with 120° angles, F atoms around central B)

The three BF bond moments cancel in vector addition.

1.55   (b) < (c) < (a).   The order is based upon hydrogen bonding with water. Compound
(b) forms no hydrogen bonds. Compound (c) does form hydrogen bonds. Compound
(a) has both hydrogen-bonding C=O and OH groups.

1.56   The two compounds are equally soluble in water because they have the same molecular weights and both form hydrogen bonds with water. Pure 1-butanol can form
hydrogen bonds, and is therefore higher boiling than diethyl ether, which cannot
form hydrogen bonds in the pure state.

1.57   (a) Lewis acid, $AlCl_3$; Lewis base, $(CH_3)_3CCl$

(b) Lewis acid, $(CH_3)_3C^+$; Lewis base, $CH_2=CH_2$

1.58   (a)   (b)   (c) CH_3OH

Chapter 2

Orbitals and Their Role in Covalent Bonding

Some Important Features

A bonding molecular orbital is formed by the overlap of two atomic or hybrid orbitals of the same phase. A bonding molecular orbital is of lower energy (more stable) than the two original orbitals. An antibonding molecular orbital is formed by interference of two atomic or hybrid orbitals of opposite phase and is of higher energy (less stable) than the two original orbitals.

A sigma (σ) molecular orbital results from the end-to-end overlap of atomic or hybrid orbitals. A pi (π) molecular orbital results from side-to-side overlap of $\underline{p}$ orbitals. A pi orbital is more exposed and is usually of higher energy (more reactive) than a sigma orbital.

A carbon atom can hybridize in one of three ways:

$\underline{sp}^3$:   four tetrahedral single bonds

$\underline{sp}^2$:   two single bonds and one double bond

$\underline{sp}$:   one single bond and one triple bond

The bond distance from a carbon atom decreases with increasing $\underline{s}$ character — that is, $\underline{sp}$ bonds are the shortest and $\underline{sp}^3$ bonds are the longest.

Some important functional groups are listed in Table 2.1 in the text (p. 56). <u>Note the different ways of writing each functional group.</u>

A nitrogen atom with three bonds to other atoms (as in NH_3, RNH_2, or $HC\equiv N$) has a pair of unshared electrons that can be donated to an electron-deficient species: $NH_3 + H^+ \longrightarrow NH_4^+$. An oxygen atom in compounds has two bonds to other atoms (as in H_2O, ROH, or $R_2C=O$) and has two pairs of unshared electrons. A carbon-oxygen or a carbon-nitrogen bond is polar because oxygen and nitrogen are more electronegative than carbon.

Conjugated double bonds can occur only if two $\underline{p}$ orbitals involved in two different pi bonds can partially overlap their sides. For this reason, the pi bonds must join adjacent atoms for conjugation to occur. Benzene (C_6H_6) is a symmetrical cyclic molecule in which six $\underline{p}$ electrons are completely delocalized (see Figure 2.23).

Resonance structures are used to show complete or partial delocalization of electronic charge, and differ only in the positions of electrons. The major contributors are the resonance structures of lowest energy (greatest stabilization). Reread the "rules" in Section 2.10C.

Reminders

Remember the valences: $C = 4$, $O = 2$, $H = 1$.

To determine the approximate bond angles around a carbon atom, ask yourself "What is the hybridization of that carbon?" If the answer is sp^3, then the bond angles are approximately 109°; if the answer is sp^2, 120°; if sp, 180°.

A doubly bonded carbon atom is sp^2 hybridized and has three planar, trigonal sigma bonds plus a pi bond.

$$X = C \underset{Z}{\overset{Y}{<}}$$

The bonds to X, Y, and Z are planar.

Answers to Problems

2.19   All are the same because C is tetrahedral.

2.20   Only (b) and (c) could be real compounds. Structure (a) has too many hydrogens for two carbons. In (a) and (d), there are odd numbers of hydrogens. (Because carbon forms four bonds, there must be an even number of hydrogens in a compound of C and H.) One way to solve a problem of this type is to write the symbols for the carbon atoms, then insert the hydrogens by trial and error. (A more sophisticated method of prediction of formulas is presented in Section 3.1B.)

2.21 sp—C—sp, two sp orbitals

2.22   (a) 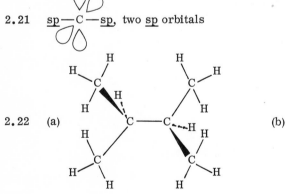   (b)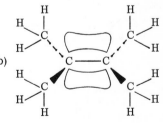

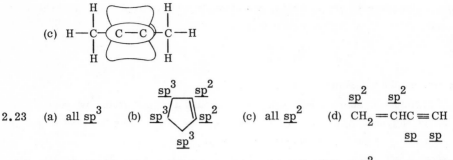

(c)

2.23   (a) all $\underline{sp}^3$ (b) $\underline{sp}^3$ $\underline{sp}^2$ ... $\underline{sp}^3$ $\underline{sp}^3$ $\underline{sp}^2$ (c) all $\underline{sp}^2$ (d) $CH_2\!=\!CHC\!\equiv\!CH$ $\underline{sp}^2$ $\underline{sp}^2$ $\underline{sp}$ $\underline{sp}$

2.24   In (a) and (b), the indicated carbon atoms are bonded to $\underline{sp}^2$ carbons and therefore lie in the same plane as the $\underline{sp}^2$ carbons. In (c), the CH_3 group bonded directly to the $\underline{sp}^2$ carbons lies in the plane of the $\underline{sp}^2$ carbons, but the other CH_3 group can rotate around its sigma bond. This CH_3 group sometimes lies in the same plane as the $\underline{sp}^2$ carbons, but is not restricted to this position.

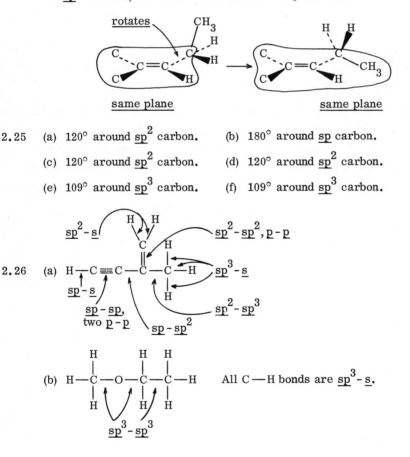

same plane   same plane

2.25   (a) 120° around $\underline{sp}^2$ carbon.   (b) 180° around $\underline{sp}$ carbon.

(c) 120° around $\underline{sp}^2$ carbon.   (d) 120° around $\underline{sp}^2$ carbon.

(e) 109° around $\underline{sp}^3$ carbon.   (f) 109° around $\underline{sp}^3$ carbon.

2.26   (a)

(b) All C—H bonds are $\underline{sp}^3$-$\underline{s}$.

(c)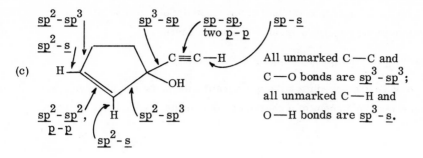

All unmarked C—C and C—O bonds are sp^3-sp^3; all unmarked C—H and O—H bonds are sp^3-s.

(d) $(CH_3)_2NH$: All C—H and N—H bonds are sp^3-s and the C—N bonds are sp^3-sp^3.

2.27 (a) 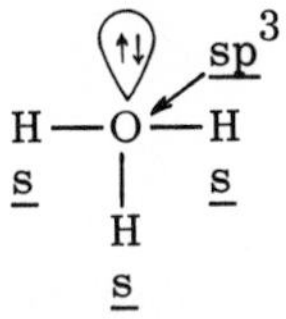 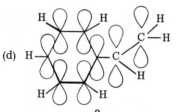 (b) 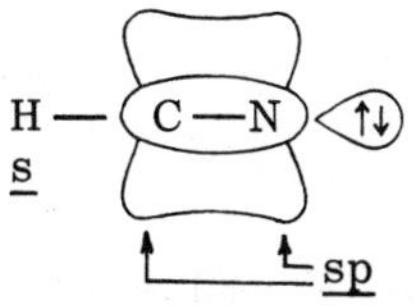 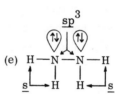 (c) 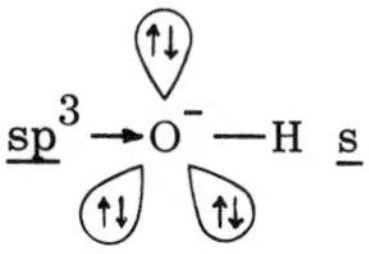 $sp^3 \rightarrow O^- $—H $\underline{s}$

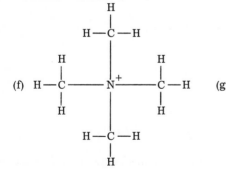

(e) H—N—N—H

Each C is sp^2, and each H is $\underline{s}$.

(f)

$$H-C-H$$

H—C————N$^+$————C—H

H—C—H

(g) $sp^3 \rightarrow O^- $—C—H $\underline{s}$

Each C and N is sp^3 and each H is $\underline{s}$.

2.28 (a)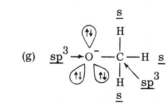

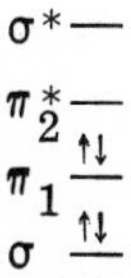

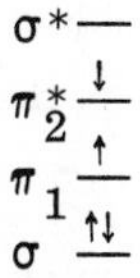

In (a), the ground state, the electrons are paired and fill the two lowest-energy orbitals. Upon excitation, an electron is promoted from the highest occupied molecular orbital (π_1) to the lowest unoccupied molecular orbital (π_2^*). Any other transition would require a greater amount of energy.

2.29 (a) 2   (b) 2   (c) 2   (d) 2

The answers are based upon the comparative amounts of s character in the bonds. The greater the amount of s character, the shorter is the bond distance.

2.30 (a) $CH_3\ddot{C}\ddot{l}$:   (b) $CH_3\ddot{N}H_2$ (c) [structure]   (d) $CH_3\overset{\ddot{O}}{\overset{||}{C}}OH$

(e) [structure]   (f) $CH_3\ddot{N}HCH_3$ (g) $CH_3\overset{\ddot{O}}{\overset{||}{C}}NH_2$

2.31 (a) CH_3CH_2OH, because an O—H bond is more polar (has a greater difference in electronegativity) than a C—O bond.

(b) $(CH_3)_2C=O$, because the electrons of the C—O pi bond are more easily polarized than those of the C—O sigma bond.

(c) CH_3NH_2, because the N—H bond is more polar than the C—N bond.

(d) CH_3CH_2Cl, because the C—Cl bond is more polar than the C—H bond.

2.32 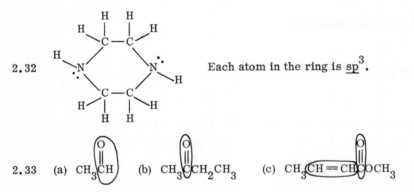 Each atom in the ring is sp^3.

2.33 (a) $CH_3\overset{O}{\overset{||}{CH}}$ (b) $CH_3\overset{O}{\overset{||}{C}}CH_2CH_3$ (c) $CH_3CH=CH\overset{O}{\overset{||}{C}}OCH_3$

In (a), the C, H, and O atoms must be circled. In some structures, it is not always easy to determine exactly which atoms to circle. In (b), if you also circled the two C atoms attached to the C=O group (as well as the C=O group), you are correct. In (c), C=C is the functional group, but it is not incorrect to include the H atoms in the circle, as we have shown. On the other hand, circling only C=O (and not the other O) in (c) is incorrect, because the functional group is an ester group, not a keto group.

2.34 (a) and (c) contain pi bonds. Write out the formula in more detail, and remember that carbon has a valence of 4; oxygen, 2; and hydrogen, 1. Note that —CHO represents an aldehyde, and —CH_2OH represents an alcohol.

(a) CH_3CH_2CHO = $CH_3CH_2\overset{\overset{\displaystyle O}{\|}}{C}H$ (b) CH_3CH_2OH = $CH_3\overset{\overset{\displaystyle H}{|}}{\underset{\underset{\displaystyle H}{|}}{C}}-O-H$

(c) $CH_3CH_2CO_2CH_3$ = $CH_3CH_2\overset{\overset{\displaystyle O}{\|}}{C}OCH_3$

2.35 (a) $\boxed{CH_2\!=\!CH}$—⬡—$\boxed{OCH_3}$ double bond, alkoxyl

(b) [furan ring]—$CH_2\boxed{OH}$ alkoxyl, hydroxyl

(c) [cyclohexene ring]—$\boxed{OH}$ double bond, hydroxyl

(d) [bicyclic lactone structure] ester   (e) $CH_3\overset{\overset{\displaystyle \boxed{NH_2}}{|}}{C}H\boxed{CO_2H}$ amino, carboxyl

(f) [steroid structure with CH_3, $\boxed{OH}$, $\boxed{HO}$] hydroxyl

(g) [steroid structure with keto, CH_3, $\boxed{OH}$, hydroxyl, $\boxed{CCH_2OH}$, keto, keto, double bond, H_3C]

2.36 (a) ROH (b) $R\overset{\overset{\displaystyle O}{\|}}{C}OR'$ or RCO_2R' (c) $RNHR'$ (d) ROR'

2.37 (a) ester   (b) ketone   (c) aldehyde   (d) ester   (e) ketone, ester

2.38 (a) $CH_3CH\!=\!CH_2$ (b) $CH_3C\!\equiv\!CH$ (c) $CH_3CH_2OCH_3$

(d) $CH_3CH_2CH_2OH$ or $(CH_3)_2CHOH$

(e) $CH_3CH_2CH_2NH_2$, $(CH_3)_2CHNH_2$, $(CH_3)_3N$, or $CH_3CH_2NHCH_3$

(f) $(CH_3)_2C\!=\!O$ (g) $CH_3CH_2\overset{\overset{\displaystyle O}{\|}}{C}H$ (h) $CH_3CH_2CO_2H$

(i) $CH_3CO_2CH_3$ or $HCO_2CH_2CH_3$

2.39   (a) $CH_3CH_2CH_2\overset{\displaystyle O}{\overset{\|}{C}}OH$ (b) ⟨O⟩$-\overset{\displaystyle O}{\overset{\|}{C}}H$ (c) ⟨O⟩$-\overset{\displaystyle O}{\overset{\|}{C}}OCH_3$

2.40   (a) ⟨O⟩$-CO_2CH_3$ (b) ⟨⟩$-OH$ (c) $CH_2\!=\!CH(CH_2)_2CH_3$

(d) CH_3CH_2CHO.   (There are other answers for all of these.)

2.41   (a), (c), and (f) are alcohols, ROH.

(b) and (h) are ethers, ROR′.

(d), (i), and (j) are alkanes, RH.

(e) and (g) are alkyl halides, RCl and RBr.

2.42   (a) $CH_2\!=\!CHCH\!=\!CHCH_3$ (b) $CH_2\!=\!CHCH_2CH\!=\!CH_2$

2.43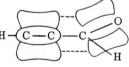

2.44   (a) equilibrium   (b) equilibrium   (c) resonance   (d) resonance

The structures in (a) and (b) differ in the positions of atoms, while the structures in (c) and (d) differ only in the positions of electrons.

2.45   (a) conjugated   (b) isolated   (c) isolated   (d) conjugated
(e) conjugated   (f) isolated.

In (e), only one of the pi bonds of the triple bond undergoes partial overlap with the $C\!=\!O$ pi bond:

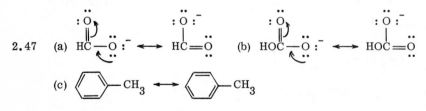

2.46   (a), (b), (d), and (f) show resonance structures – structures that differ only in the positions of electrons.

2.47   (a) $H\overset{\displaystyle :O}{\overset{\|}{C}}\!-\!\overset{..}{\underset{..}{O}}:^- \longleftrightarrow H\overset{\displaystyle :O:^-}{\overset{|}{C}}\!=\!\overset{..}{\underset{..}{O}}$ (b) $HO\overset{\displaystyle :O}{\overset{\|}{C}}\!-\!\overset{..}{\underset{..}{O}}:^- \longleftrightarrow HO\overset{\displaystyle :O:^-}{\overset{|}{C}}\!=\!\overset{..}{\underset{..}{O}}$

(c) ⟨O⟩$-CH_3 \longleftrightarrow$ ⟨O⟩$-CH_3$

2.48   (a), because it shows each C and N with an octet and has no charge separation.

2.49   (a)

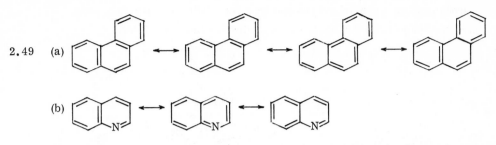

(b)

2.50   The third one in each case. In (a), because the third structure has a charge separation. In (b), because the nitrogen has only six bonding electrons. In (c), because the carbonyl group is polarized in the opposite direction of the relative electronegativities.

2.51   (a) CH_2=C=CH_2 (b)

$\underline{sp}^2$ $\underline{sp}$ $\underline{sp}^2$

(c) No, they cannot overlap their sides.

(d) No, two pi bonds cannot delocalize electronic charge without overlap.

2.52   :O :: C :: O:

With a bond angle of 180° and two pi bonds, the carbon atom must be $\underline{sp}$ hybridized.

2.53   (a) $CH_3\overset{..}{N}H$—$\overset{\overset{+}{N}H_2}{\overset{\|}{C}}$—$NH_2$ ⟷ $CH_3\overset{..}{N}H$—$\overset{:NH_2}{\underset{+}{C}}$=$NH_2$ ⟷ $CH_3\overset{+}{N}H$=$\overset{:NH_2}{C}$—NH_2

(b) $CH_3\overset{\overset{+}{N}H_2}{\overset{\|}{C}}$—$NH_2$ ⟷ $CH_3\overset{:NH_2}{C}$=$\underset{+}{N}H_2$

2.54 H_2^+, or H · H$^+$,   σ* ___

                                                              σ —↑—

        H_2^-, or H : H$^-$,   σ* —↑—

                                                              σ —↑↓—

The H_2^+ ion contains one less electron than H_2 — that is, the group contains one electron. The H_2^- ion contains one more electron than H_2. Since the sigma bonding orbital in H_2^- is filled, the third electron must be in the antibonding orbital.

2.55   Carbon is in the <u>sp</u> state and has two <u>sp-s</u> bonds with the hydrogen atoms. The
carbon atom has two <u>p</u> orbitals with one electron in each orbital.

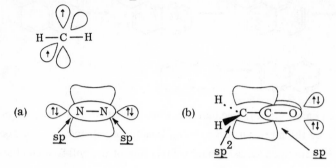

2.56   (a)   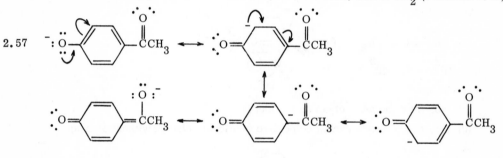 (b)

The structure in (b) is similar to allene (Problem 2.51) and to CO_2 (Problem 2.52).

2.57

Chapter 3

Structural Isomerism, Nomenclature, and Alkanes

Some Important Features

Structural isomers are compounds with the same molecular formulas but different structures (orders of attachment of atoms).

A saturated open-chain alkane (branched or not) has the general formula C_nH_{2n+2}. We subtract two hydrogen atoms from the general formula for each double bond or for each ring.

We will not reiterate the nomenclature of organic compounds here, but refer you to Chapter 3 in the text. (For quick reference, see the appendix in the text.) Remember to: (1) number a chain or ring to give the lowest number to the principal functional group, and (2) group like substituents together.

Alkanes are generally nonreactive, but they do undergo combustion and reaction with halogens. The heat of combustion for alkanes increases with increasing molecular weight and also with increasing energy contained in the bonds.

Reminders

The same molecule may be drawn in a number of ways:

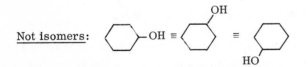

To be structural isomers, two molecules must have the atoms attached in different orders:

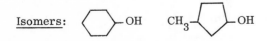

Some instructors suggest naming potential isomers; if the names are the same, the structures represent the same compound. (This is a valid approach only if you name both structures correctly.)

When drawing isomers for a particular molecular formula, use the general formula. If a compound contains a ring, a structural isomer can contain a double bond instead.

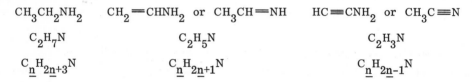 and $CH_2=CHCH_3$:   both C_nH_{2n}

General formulas may be extrapolated to compounds other than hydrocarbons:

$$CH_3CH_2CH_2OH = C_nH_{2n+2}O$$
$$\uparrow \text{no double bond or ring}$$

$$\overset{\displaystyle O}{\overset{\|}{CH_3CH_2CH}} = C_nH_{2n}O$$
$$\uparrow \text{— one double bond or ring}$$

Because nitrogen has a valence of 3, take special care in using general formulas with nitrogen compounds.

$CH_3CH_2NH_2$	$CH_2=CHNH_2$ or $CH_3CH=NH$	$HC\equiv CNH_2$ or $CH_3C\equiv N$
C_2H_7N	C_2H_5N	C_2H_3N
$C_nH_{2n+3}N$	$C_nH_{2n+1}N$	$C_nH_{2n-1}N$

Answers to Problems

3.12   (b), (c), (d), (f), (g)

3.13   (a) $CH_3CH_2CH_2OH$, $CH_3OCH_2CH_3$, $(CH_3)_2CHOH$

(b) $CH_3CH_2CH_2Cl$, $(CH_3)_2CHCl$

3.14   (c), (d), (e), and (f) have no structural isomers. A typical structural isomer for (a) would be $CH_2=CHCH_2OH$, and for (b), CH_3CHBr_2

3.15   (c) and (f) are isomers. In (a), the formulas represent a five-carbon and a six-carbon alcohol. In (d) and (e), the formulas represent the same compounds.

3.16   (a) $CH_3(CH_2)_4CH_3$, $(CH_3)_2CHCH_2CH_2CH_3$, $(CH_3)_3CCH_2CH_3$

$(CH_3)_2CHCH(CH_3)_2$, $CH_3CH_2\overset{\displaystyle CH_3}{\overset{|}{CH}}CH_2CH_3$

(b) $CH_3CH_2CH_2CH_2OH$, $(CH_3)_2CHCH_2OH$, $CH_3CH_2\overset{\displaystyle CH_3}{\overset{|}{CH}}OH$, $(CH_3)_3COH$

(c) $CH_3CH_2CH_2CH_2NH_2$, $(CH_3)_2CHCH_2NH_2$, $CH_3CH_2\overset{\underset{\displaystyle CH_3}{|}}{C}HNH_2$, $(CH_3)_3CNH_2$,

$CH_3CH_2CH_2NHCH_3$, $(CH_3)_2CHNHCH_3$, $(CH_3CH_2)_2NH$, $CH_3CH_2N(CH_3)_2$

(d) $CH_3CH_2CHBrCl$, $CH_3CHBrCH_2Cl$, $BrCH_2CH_2CH_2Cl$, $CH_3CHClCH_2Br$,

$(CH_3)_2CBrCl$

(e) $CH_3CH_2C\equiv CH$, $CH_3C\equiv CCH_3$, $CH_2=CHCH=CH_2$, $CH_2=C=CHCH_3$,

☐ , CH_3—◁ , CH_3—◁ , CH_2=◁

3.17　(a) and (f) are isomers;　(b) and (e) are the same compound.

3.18　(a) C_6H_{12}, $C_{\underline{n}}H_{2\underline{n}}$　(b) C_8H_{14}, $C_{\underline{n}}H_{2\underline{n}-2}$

(c) C_8H_{14}, $C_{\underline{n}}H_{2\underline{n}-2}$　(d) $C_{14}H_{24}$, $C_{\underline{n}}H_{2\underline{n}-4}$

3.19　(a) $CH_3CH=CH_2$　(b)　(c)　(d)

(There are other correct answers.)

3.20　(a) $CH_3CH_2C\equiv CCH_2CH_3$　(b) $CH_3\overset{\underset{\displaystyle ||}{O}}{C}OH$　(c) $CH_3\overset{\underset{\displaystyle ||}{O}}{C}CH_3$　(d) △

3.21　CH_3OH, CH_3CH_2OH, $CH_3CH_2CH_2OH$, $CH_3CH_2CH_2CH_2OH$, $CH_3(CH_2)_4OH$

3.22　$CH_3(CH_2)_4Br$, $CH_3(CH_2)_5Br$, $CH_3(CH_2)_6Br$, $CH_3(CH_2)_7Br$, $CH_3(CH_2)_8Br$,

$CH_3(CH_2)_9Br$

3.23　(a) $(CH_3)_3C(CH_2)_5CH_3$

(b) $(CH_3CH_2)_2CH\overset{\underset{\displaystyle CH_2CH_3}{|}}{C}HCH_2CH_2CH_3$

(c) $(CH_3)_2CHCH_2\overset{\underset{\displaystyle CH_2CH_3}{\overset{\displaystyle CH_3}{|}}}{C}(CH_2)_4CH_3$

(d)

(e) $CH_3CH_2\overset{\underset{\displaystyle CH_3}{|}}{C}H$—

(f) $(CH_3)_3C$—

(g) $(CH_3)_2CHCH_2$—

(h) $CH_3(CH_2)_4$—

(i) $(CH_3CH_2CH_2)_2CHCH(CH_3)_2$

3.24   (a) 3-ethylpentane   (b) 2,3-dimethylbutane

   (c) 5-ethyl-1,2,3-trimethylcyclohexane   (d) <u>t</u>-butylcyclopentane

   (e) 2,2,4,4-tetramethylheptane

   Be sure that the parent carbon chain is numbered from the end that will give the smallest numbers. In (d), no number is necessary for a monosubstituted cyclo-alkane.

3.25   (a) 3-ethylhexane; the longest chain was not chosen as the parent.

   (b) 3-methyloctane; the parent was not numbered to give the smallest numbers.

   (c) 2,2-dimethylpropane; same as (a).

   (d) 3-ethyl-3,4-dimethyldecane; the methyl groups were not grouped together.

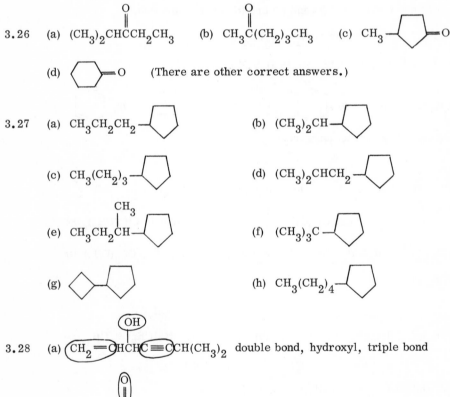

3.26   (a) $(CH_3)_2CHCCH_2CH_3$ (b) $CH_3C(CH_2)_3CH_3$ (c) CH_3—cyclopentane=O

   (d) cyclohexane=O   (There are other correct answers.)

3.27   (a) $CH_3CH_2CH_2$—cyclopentane   (b) $(CH_3)_2CH$—cyclopentane

   (c) $CH_3(CH_2)_3$—cyclopentane   (d) $(CH_3)_2CHCH_2$—cyclopentane

   (e) $CH_3CH_2CH(CH_3)$—cyclopentane   (f) $(CH_3)_3C$—cyclopentane

   (g) cyclopropyl-cyclopentane   (h) $CH_3(CH_2)_4$—cyclopentane

3.28   (a) CH_2=$CHCHC$≡$CH(CH_3)_2$ with OH   double bond, hydroxyl, triple bond

   (b) CH_3CH_2 keto, alkoxyl (ether)

   Note that (b) is not an ester because the oxygen is not bonded directly to the carbonyl carbon.

   (c) $BrCH_2CCO_2H$ bromo, keto, carboxyl

(d)   ester, aldehydo, triple bond

3.29 $(CH_3)_2CHCH_2C(CH_3)_3$, $(CH_3)_2\overset{\overset{\displaystyle CH_3}{|}}{C}HCHCH(CH_3)_2$, $(CH_3)_2CH\overset{\overset{\displaystyle CH_3}{|}}{\underset{\underset{\displaystyle CH_3}{|}}{C}}CH_2CH_3$

$CH_3CH_2\overset{\overset{\displaystyle CH_2CH_3}{|}}{\underset{\underset{\displaystyle CH_3}{|}}{C}}CH_2CH_3$, $(CH_3)_2CH\overset{\overset{\displaystyle CH_2CH_3}{|}}{C}HCH_2CH_3$, $(CH_3)_3\overset{\overset{\displaystyle CH_3}{|}}{C}CHCH_2CH_3$

3.30   (a) $CH_3(CH_2)_4Cl$, $CH_3CHCl(CH_2)_2CH_3$, $CH_3CH_2CHClCH_2CH_3$

(b) — Cl   (c) $ClCH_2\overset{\overset{\displaystyle CH_3}{|}}{\underset{\underset{\displaystyle CH_3}{|}}{C}}CH_2CH_3$, $(CH_3)_3CCHClCH_3$, $(CH_3)_3CCH_2CH_2Cl$

(d) $(CH_3)_3CCH_2Cl$

3.31   (a) 1,2,3,4,5,6-hexachlorocyclohexane   (b) 1,1,2-trichloroethane

(c) 2-bromo-3-chloro-2,3-dimethylbutane   (d) nitromethane

(e) 5-bromo-1,1,1-trichloro-5-ethyl-2,3-dimethylheptane

3.32   (a) $Br\overset{\overset{\displaystyle C_6H_5}{|}}{\underset{\underset{\displaystyle C_6H_5}{|}}{C}}HCHCH_3$ (b) CCl_3CCl_3 (c) $HOCH_2CHI(CH_2)_5CH_3$ (d)

3.33   (a) 2-pentene   (b) 2-methyl-2-butene   (c) 1,3-cyclopentadiene

(d) 3-methyl-1,3,5-hexatriene   (e) 2-butyne   (f) 1,1,2-trichloroethene

(g) 3-bromo-1-cyclohexene

3.34   (a) $CH_3(CH_2)_5CHO$, heptanal   (b) Cl_2CHCO_2H, dichloroethanoic acid

(c) $CH_3(CH_2)_3\overset{\overset{\displaystyle O}{\|}}{C}(CH_2)_3CH_3$, 5-nonanone

3.35   (a) 2-methylpropanoic acid   (b) ethyl 2-chloropropanoate

3.36   (a) $CH_3CH_2CO_2CH(CH_3)_2$ (b) $CH_3(CH_2)_3\overset{\overset{\displaystyle CH_3CH_2}{|}}{\underset{\underset{\displaystyle CH_2CH_3}{|}}{C}}\overset{\overset{\displaystyle CH_3}{|}}{C}HCH_2CHO$ (c)

3.37   The heat of combustion is approximated by using the value for butane (688 kcal/mole) and adding the heat of combustion of four CH_2 groups (157 kcal/mole each).

$$688 + 4(157) = \text{approximately } 1316 \text{ kcal/mole}$$

3.38   (a) hexane   (b) 2-butene   (c) 1-pentanol

Branching decreases the boiling point by interfering with van der Waals attractions.

3.39   (a) Small droplets of water (insoluble in gasoline) freeze and clog the carburetor or other portions of the gas line.

(b) Methanol (soluble in gasoline) solubilizes the water in the gasoline.

3.40   Cyclopropane is more efficient because it has a higher heat of combustion per CH_2 group.

3.41 $CH_3CH_2\overset{\overset{\displaystyle O}{\|}}{C}CH_3$ Because the general formula is $C_nH_{2n}O$, there must be one site of unsaturation or one ring. Since the compound is a ketone, there can be no carbon-carbon double bond nor a ring. Only one four-carbon, open-chain ketone fits the formula.

3.42

The general formula is $C_nH_{2n}O$; therefore, there must be either one site of unsaturation or one ring. Since the problem states that the compound is an alcohol and contains no carbon-carbon double bonds, the structure must contain either a three- or four-membered ring.

3.43 $CH_3CH_2CH_2CO_2H$, $(CH_3)_2CHCO_2H$

Again, the general formula is $C_nH_{2n}O$. The unsaturation must occur in the $C{=}O$ of the carboxyl group; therefore, the carboxylic acid must be open-chain.

3.44 $H\overset{\overset{\displaystyle O}{\|}}{C}CH_2CH_2CO_2H$, $H\overset{\overset{\displaystyle O}{\|}}{C}\underset{\underset{\displaystyle CH_3}{|}}{C}HCO_2H$

Both the aldehyde and the carboxyl groups must be at the ends of a chain. Thus, the other two carbon atoms must be positioned between these two functional groups.

3.45 $\quad \underset{\text{CH}_3\overset{\text{O}}{\overset{\|}{\text{C}}}\text{CH}_2\overset{\text{O}}{\overset{\|}{\text{C}}}\text{H},}{} \quad \underset{\text{CH}_3\text{CH}_2\overset{\text{O}}{\overset{\|}{\text{C}}}\underset{\overset{\|}{\text{O}}}{\text{C}}\text{H},}{} \quad \underset{\text{H}\overset{\text{O}}{\overset{\|}{\text{C}}}\text{CH}_2\text{CH}_2\overset{\text{O}}{\overset{\|}{\text{C}}}\text{H},}{} \quad \underset{\text{H}\overset{\text{O}}{\overset{\|}{\text{C}}}\text{CH}\overset{\text{O}}{\overset{\|}{\text{C}}}\text{H},}{\underset{\overset{|}{\text{CH}_3}}{}} \quad \underset{\text{CH}_3\overset{\text{O}}{\overset{\|}{\text{C}}}-\overset{\text{O}}{\overset{\|}{\text{C}}}\text{CH}_3}{}$

One oxygen is in a carbonyl group, leaving one oxygen to be determined. Since the second oxygen is not in a hydroxyl, ether, ester, or carboxyl grouping, we conclude that the second oxygen is also part of a carbonyl group and that both carbonyl groups are aldehyde or keto groups. The general formula $C_nH_{2n-2}O_2$ allows for two sites of unsaturation ($C{=}O$ and $C{=}O$); therefore, no ring can be present.

Chapter 4

Stereochemistry

Some Important Features

Geometric (or <u>cis</u>, <u>trans</u>) isomerism arises from attachments being on the same side or on opposite sides of a double bond or ring.

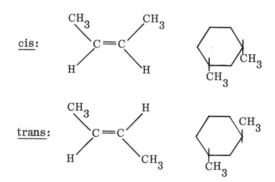

Geometric isomerism is not possible around a double bond if two groups on one carbon atom are the same (two CH_3 groups in the following example).

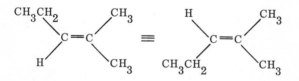

The above two formulas do not represent isomers because they are superimposable. Any superimposable molecules represent the same compound, not isomers. If this feature is not clear to you, make molecular models of the above two formulas and prove to yourself that they are the same.

The priority rules for the (<u>E</u>) and (<u>Z</u>) system are listed in Section 4.1B. The most important feature is that <u>higher atomic number</u> means higher priority. (In the case of isotopes, <u>higher atomic mass</u> means higher priority.) If the atoms attached to the sp^2 carbons are the same, proceed along the chain to the first point of difference.

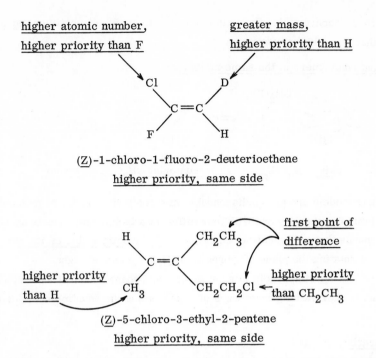

(Z)-1-chloro-1-fluoro-2-deuterioethene

higher priority, same side

(Z)-5-chloro-3-ethyl-2-pentene

higher priority, same side

A common student error is to sum atomic numbers to arrive at priorities of groups. This is incorrect.   Look at individual atoms at the first point of difference, not at groups of atoms.

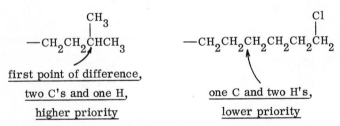

Conformations of molecules are the different shapes they can assume. Anti conformations are generally more stable than eclipsed conformations. (Compare the following Newman projections with molecular models.)

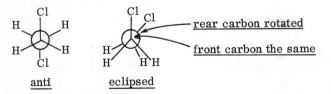

anti   eclipsed

Conformations of cyclic compounds (six-membered rings particularly) should be studied carefully. A molecular model of cyclohexane shows ring conformation much better than a "paper formula" does.

Attachments to the cyclohexane ring in the chair form can be equatorial or axial. (The axial substituents are especially easy to see with models.) Although the following conformers are interconvertible, the conformer with the bulkier substituent <u>equatorial</u> is more stable (that is, of lower energy).

<u>Different conformers of the same molecule:</u>

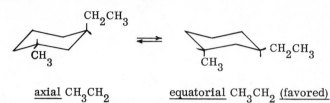

<u>axial</u> CH_3CH_2 <u>equatorial</u> CH_3CH_2 (favored)

Molecular models are also indispensable in a study of chirality of molecules. A chiral carbon atom is generally one with four different attachments. The presence of one chiral carbon means that a molecule is chiral, or <u>not superimposable on its mirror image</u>, and is capable of rotating the plane of vibration of plane-polarized light.

A pair of nonsuperimposable mirror images are enantiomers of each other. The following formulas show two different ways of representing the enantiomers of 2-chloro-butane.

chiral carbon   mirror

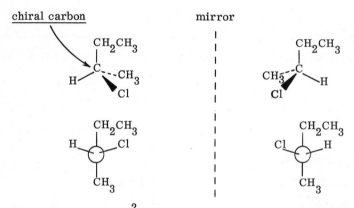

If a carbon atom is sp^2-hybridized, or if it has two identical attachments, the carbon atom is <u>achiral</u> (not chiral). (The chirality of a molecule as a whole, however, is determined by the lack of superimposability on the mirror image.) The following two examples are superimposable on their mirror images. (Try it with models.)

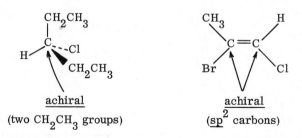

<u>achiral</u>   <u>achiral</u>
(two CH_2CH_3 groups)   (sp^2 carbons)

If a molecule contains more than one chiral carbon, there is a possibility of an internal plane of symmetry. A molecule with chiral carbons but with an internal plane of symmetry is the <u>meso</u> form of the compound and cannot rotate the plane of vibration of plane-polarized light.

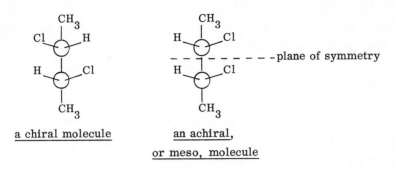

a chiral molecule   an achiral,
                              or meso, molecule

A <u>meso</u> molecule has a "top half" that is the mirror-reflective image of the "bottom half." Because groups can rotate around their bonds, it is not always easy to see at a glance if a structure is a <u>meso</u> form. Molecular models are useful here. Construct the models of the preceding two compounds and their mirror-reflective images, and verify the chirality or achirality of the two. Then, rotate groups around the bonds to see the different conformations each molecule can assume.

The (<u>R</u>) and (<u>S</u>) system of assigning the absolute configuration around chiral carbon atoms is discussed in Section 4.8A. You will probably find assignment of configuration easiest to do with models.

Reminders

To draw a chair form of cyclohexane quickly, follow the steps below (or purchase a template).

(1)   (2)   (3)

In drawing equatorial and axial substituents, first draw the axial bonds, which are vertical.

Now draw the equatorial bonds, keeping the bond angles at approximately 109°.

To determine <u>cis</u> and <u>trans</u> on a ring, decide if the groups are on the same side or on opposite sides.

cis (both on "top")

To draw the enantiomer of a ball-and-stick formula, simply reverse two groups, usually those attached to the "front" bonds (B, below).

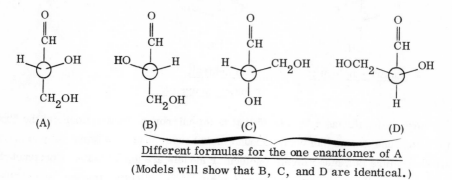

(A)   (B)   (C)   (D)

Different formulas for the one enantiomer of A

(Models will show that B, C, and D are identical.)

Answers to Problems

4.16   (a) $CH_3CH_2CH_2CH_2CH=CH_2$, no geometric isomer because there are two identical groups (H) on one carbon of the double bond.

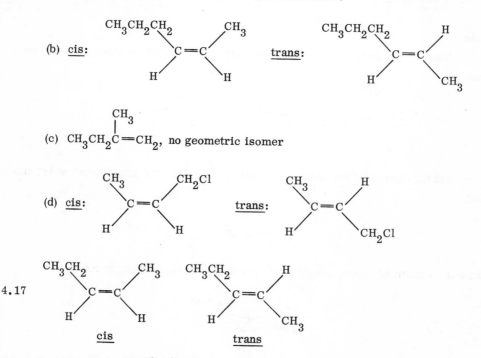

(b) cis: trans:

(c) $CH_3CH_2C=CH_2$, no geometric isomer

(d) cis: trans:

4.17

cis   trans

1-Pentene, 2-methyl-1-butene, and 2-methyl-2-butene do not have geometric isomers.

4.18 $CH_3CH_2CH=CH_2$ $(CH_3)_2C=CH_2$

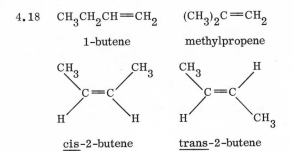

cis-2-butene   trans-2-butene

4.19   Compounds (a) $C_6H_5CH=CHC_6H_5$, (c) $CH_3CH=CHC\equiv CH$, and (e) have geometric isomers. In (c), the isomerism occurs around the double bond, not around
the triple bond, which is linear. Compounds (b) $CH_2=CHC\equiv CH$ and (d)
$(CH_3)_2C=CCH_2CH_3$ do not have geometric isomers because one carbon has two
$\overset{|}{C}H_3$
identical groups in each case.

4.20   In each case, the higher-priority atoms or groups are circled.

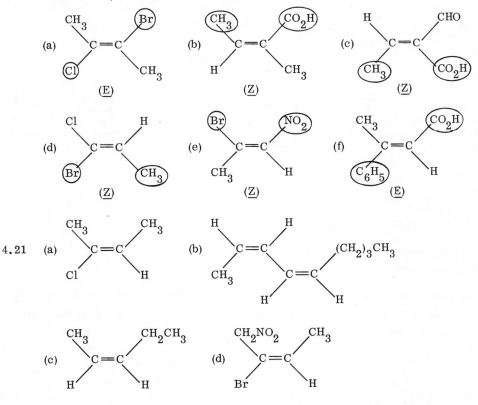

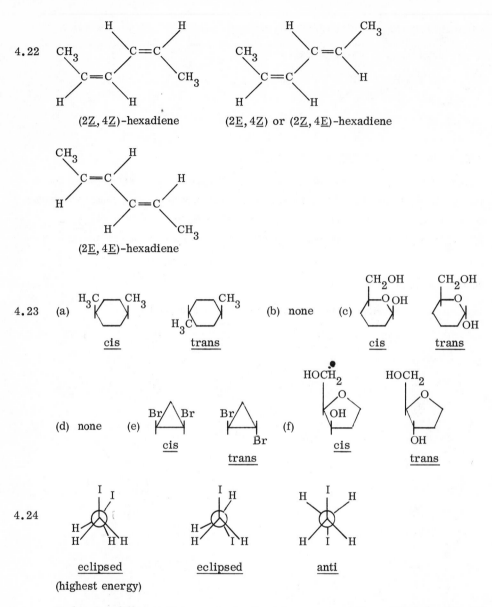

4.22

(2Z, 4Z)-hexadiene

(2E, 4Z) or (2Z, 4E)-hexadiene

(2E, 4E)-hexadiene

4.23   (a)

cis   trans   (b) none   (c)   cis   trans

(d) none   (e)   cis   trans   (f)   cis   trans

4.24

eclipsed   eclipsed   anti

(highest energy)

In drawing different Newman projections for a compound, rotate only one of the two carbons of the projection. In our answer, the front carbon is fixed and the rear carbon is rotated.

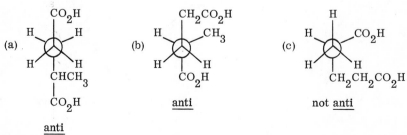

4.25   (a)   (b)   (c)

anti   anti   not anti

In (a) and (b), the two largest groups are anti. There is no anti-conformation for (c) around the two carbons shown because both large groups are on the same carbon.

4.26   (a) e   (b) a   (c) e   (d) a

4.27   A boat conformation is of higher energy than a chair form; therefore, (a) and (b) are of higher energy than (c) or (d). The conformation in (d) has two axial groups, while the one in (c) has one axial and one equatorial. Therefore, (c) is the lowest-energy, most-stable conformation.

4.28   (a), because the large t-butyl group is equatorial.

4.29   (a) In the cis-isomer, both substituents can be equatorial.

   (b) No, because any conformation of the cis-isomer has one group e and the other group a.

4.30

4.31

The most stable conformer is the one in which the largest group (propyl) is in an equatorial position.

4.32

e, e

4.33   (a) $(CH_3)_2CCH_2CH_3$

   with C_6H_5 attached

achiral because each sp^3 carbon has at least two identical attachments

   (b) $C_6H_5CH-CCH_3$ with Br, Br and C_6H_5

chiral carbons

4.34   (a) $(CH_3)_2CH\overset{*}{C}HBrCH_3$ (b) $CH_3CH_2CH_2\overset{*}{C}HOH$ with CH_3

   (c) $H_2N\overset{*}{C}HCO_2H$ with CH_3 (d) $CH_3\overset{*}{C}HBr\overset{*}{C}HBrCH_2CH_3$ (e) none

4.35   In each case, the plane is perpendicular to the ring at the dashed line. (If you have difficulty visualizing the planes, make molecular models.)

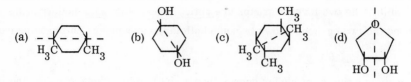

4.36   The planes are perpendicular to the rings at the dashed lines.

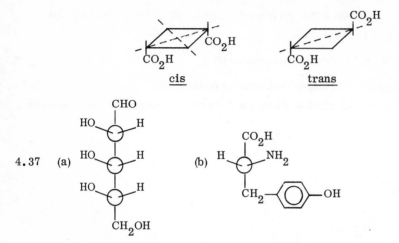

4.37   (a)   (b)

The easiest way to solve this type of problem is to draw a line representing a mirror and then draw the mirror reflection at each chiral carbon. The result is that each group originally on the left is switched to the right and vice versa. (We need not reverse groups on achiral carbon atoms because the configuration is the same regardless of how we draw these groups.)

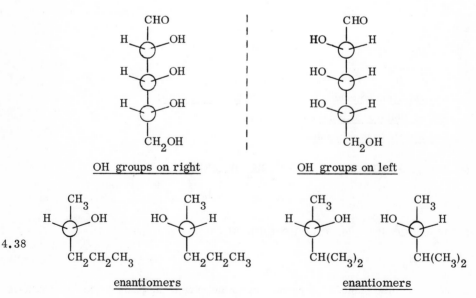

4.39 $\underset{\underset{\text{Br}}{|}}{\text{CH}_3\text{CHCO}_2\text{H}}$, not $\text{BrCH}_2\text{CH}_2\text{CO}_2\text{H}$, because the optically active compound must have a chiral carbon atom.

4.40 $\underset{\underset{\text{OH}}{|}}{\text{CH}_3\text{CH}_2\text{CHCH}_3}$

4.41   (a) 2   (b) none, (1) is the same as (b)   (c) none
        (d) none, (4) is the same as (d).

4.42   (b) and (d). The structures in (a) differ in the projection of only one chiral carbon. The structures in (c) are identical.

4.43   −13.5°

4.44   (b) and (c); (a) does not have a chiral carbon atom.

4.45   (a)   (b) none   (c)

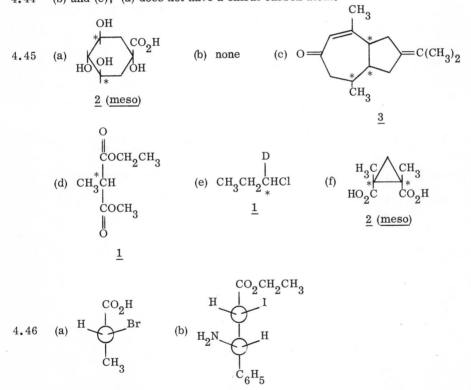

First draw the structure disregarding chirality. Next, draw the structure in a ball-and-stick formula, leaving the two horizontal bonds without groups. Finally, taking one chiral carbon at a time, insert the groups at the horizontal bonds in their proper configuration. (If your answers do not agree with those given, make models of both your answers and those answers given to compare them.)

4.47   (a) four (two chiral carbons)
        (b) two (one chiral carbon)

4.48   (a) I and III are enantiomers (nonsuperimposable mirror images).

(b) I and II are diastereomers, as are II and III (stereoisomers that are not enantiomers).

(c) II is a meso compound; a horizontal plane of symmetry may be drawn through carbon 3.

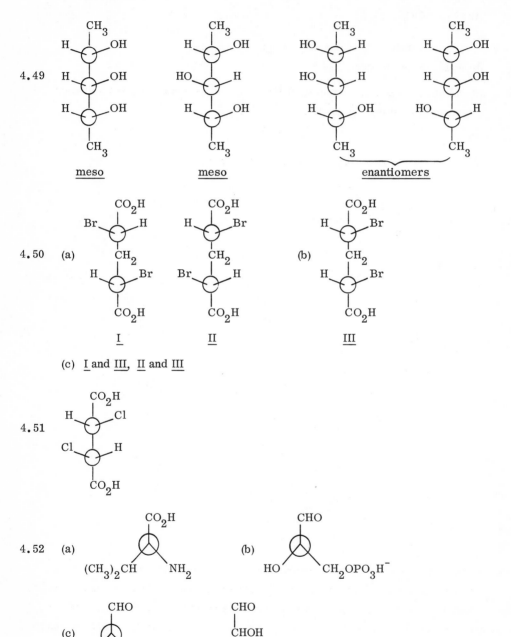

4.49

4.50   (a)

(b)

(c) I and III, II and III

4.51

4.52   (a)

(b)

(c)

4.53   (a) (R)   (b) (R)   (c) (S)

4.54   (a) (S)   (b) (S)   (c) (S)

4.55   (a) three:
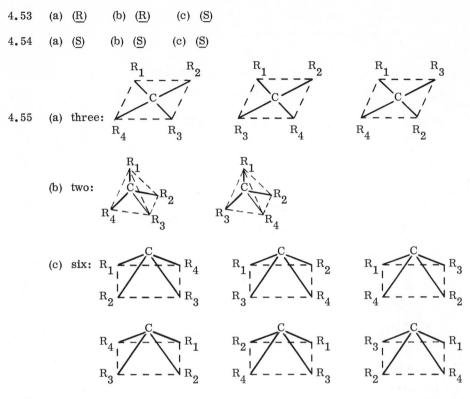

(b) two:

(c) six:

4.56   (a) none   (b) one pair   (c) three pairs (shown one above the other).

4.57
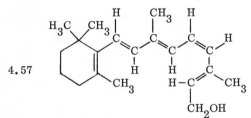

4.58   none

4.59   (a)
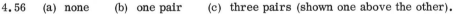
CH$_2$OH   (b) HOCH$_2$ O

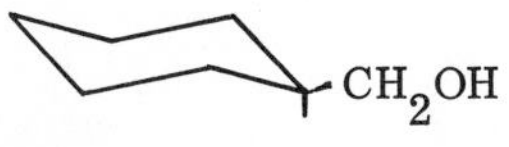

OH

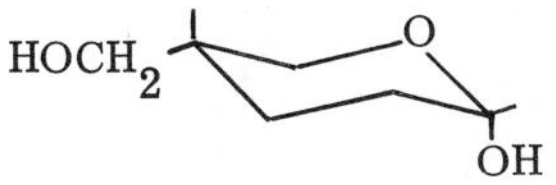

(c)

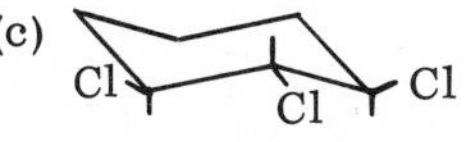

Cl   Cl   Cl

In (b), the bulkier group is equatorial in the most stable conformation.

4.60   (a) H$_3$C   CH(CH$_3$)$_2$ (b)   CH$_3$ Cl

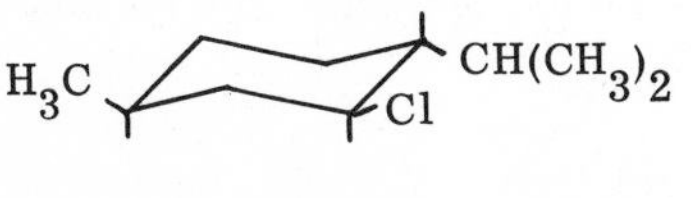

Cl   (H)   CH(CH$_3$)$_2$

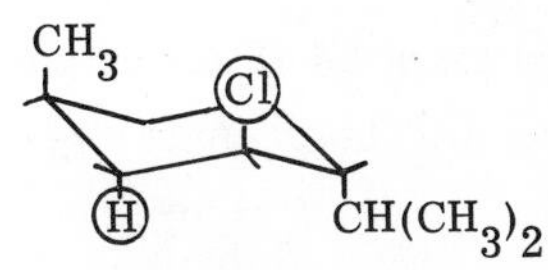

not favored

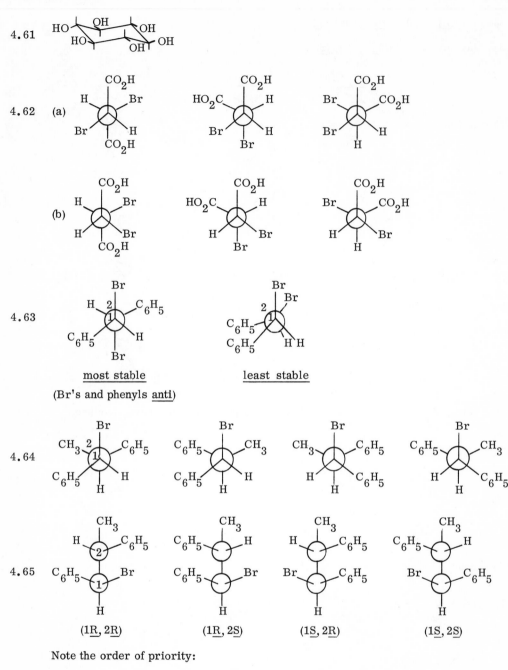

4.61

4.62 (a)

(b)

4.63

most stable

(Br's and phenyls anti)

least stable

4.64

4.65

(1R, 2R)   (1R, 2S)   (1S, 2R)   (1S, 2S)

Note the order of priority:

$$-Br > -CHBr > -C_6H_5 > -CHC_6H_5 > CH_3 > H$$
$$\qquad\quad\; \underset{C_6H_5}{|} \qquad\qquad\qquad\; \underset{CH_3}{|}$$

The phenyl group is of higher priority than the group to its right because of its
attachments. (The priority equivalents follow.)

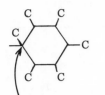 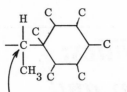

three C's attached   one H and two C's attached

4.66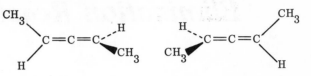

p orbitals   nonsuperimposable mirror images, or enantiomers

4.67   (a) zero, because it is a racemic mixture.

   (b) The solution is, in effect, half racemic and half (S)-enantiomer. The observed
       rotation is therefore half that of the pure (S)-enantiomer, or +8.0°.

4.68   (a) $[\alpha] = \dfrac{\alpha}{l\,c} = \dfrac{+0.45°}{0.10 \text{ dm} \times 0.200 \text{ g/ml}} = +22.5°$

   (b) $[\alpha] = \dfrac{-3.2°}{1.0 \text{ dm} \times 0.10 \text{ g/ml}} = -32°$

4.69   (b), (c), and (e) would cause rotation. Compound (a) is meso and mixture (d) is
       racemic; neither of these would cause rotation.

4.70   In (a), the mixture is 50% optically active compound plus 50% optically inactive
       compound; therefore, the rotation is half that of the optically active compound,
       or -6°. In (b), the mixture is racemic; the specific rotation is therefore zero.

4.71   (a), (c), (d), (e).

Chapter 5

Alkyl Halides;
Sustitution and
Elimination Reactions

Some Important Features

Alkyl halides, but not aryl halides, can undergo substitution or elimination when treated with a nucleophile. Primary alkyl halides and, to an extent, secondary alkyl halides generally react by an S_N2 path. (Tertiary alkyl halides do not undergo S_N2 reaction.) Inversion of configuration at the functional carbon is observed in an S_N2 reaction if that carbon is chiral.

$$S_N2: \ Nu^- + R-X \longrightarrow Nu-R + X^-$$
$$1°$$
$$(and \ 2°)$$

When treated with a very weak nucleophile, such as H_2O, ROH, or RCO_2^-, 2° and 3° alkyl halides can react by an S_N1 path. Primary alkyl halides (unless allyl or benzyl) do not undergo this reaction.

$$S_N1: \ R-X \xrightarrow{-X^-} [R^+] \xrightarrow{H_2O} \left[R\overset{+}{O}H_2\right] \xrightarrow{-H^+} ROH$$
$$3°$$
$$(and \ 2°)$$

Carbocation reactions have some disadvantages:

1. Racemization occurs because a carbocation is planar and therefore achiral.
2. Rearrangement occurs if a more stable carbocation can be formed by a 1,2-shift. Therefore, a 2° carbocation will rearrange (if possible) to a 3° carbocation. A 3° carbocation will rearrange (if possible) to an allyl or benzyl carbocation.

$$(CH_3)_2\overset{+}{C}-\overset{\overset{\displaystyle H}{|}}{C}HCH_3 \longrightarrow (CH_3)_2\overset{+}{C}-\overset{\overset{\displaystyle H}{|}}{C}HCH_3$$
$$2° \qquad\qquad\qquad 3°$$

Elimination reactions (E1 or E2) are favored by a 3° (or 2°) alkyl halide and a strong base. The principal product is usually the more-substituted, <u>trans</u> alkene.

on adjacent carbons

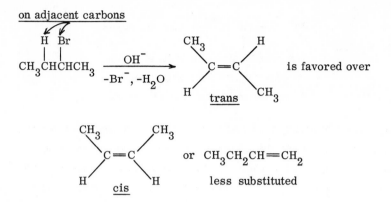

is favored over

or $CH_3CH_2CH=CH_2$
less substituted

An E1 reaction proceeds by way of a carbocation:

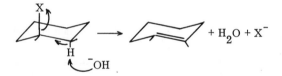

E1: $R_2CHCR_2 \xrightarrow{-Br^-} \left[R_2CH\overset{+}{C}R_2 \right] \xrightarrow[-H_2O]{OH^-} R_2C=CR_2$

An E2 reaction proceeds by $\underline{anti}$-elimination of H^+ and X^- on adjacent carbons. If the halogen atom is attached to a ring carbon, the H and X must be $\underline{trans}$ and diaxial in order to be eliminated.

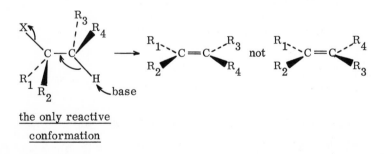

A properly selected alkyl halide yields a single geometric isomer in an E2 reaction.

the only reactive
conformation

Every reaction step proceeds through a transition state. A low-energy (more stable) transition state means a low E_{act} and therefore a fast rate of reaction.

In an S_N1 reaction, the rate-determining (slower) step is the formation of the carbocation. Any feature that stabilizes the carbocation also stabilizes the transition state and leads to a faster reaction.

$$RX \xrightarrow{-X^-} [R^+] \xrightarrow{Nu^-} RNu \text{ or alkene}$$

The rate is increased if R^+ is
stabilized by the inductive effect,
resonance, or a highly polar solvent.

Other important topics discussed in this chapter are how a carbocation is stabilized; first- and second-order rates; the kinetic isotope effect; the effects of steric hindrance on the reactivity of alkyl halides and on the products of E2 reactions (Saytseff versus Hofmann products); and how to plan a synthesis reaction. You may also want to review in Chapter 2 the rules for writing resonance structures.

Reminders

Although S_N2 reactions yield chiral products from chiral reactants, S_N1, E1, and E2 reactions all lead to racemization at the functional carbon.

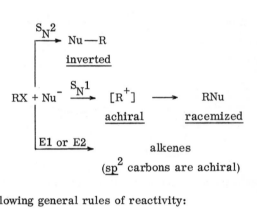

Remember the following general rules of reactivity:

S_N2: $CH_3X > 1° \ RX > 2° \ RX$
S_N1: $3° \ RX > 2° \ RX$
E2: $3° \ RX > 2° \ RX > 1° \ RX$
E1: $3° \ RX > 2° \ RX$

Strong Nu^-, high concentration: S_N2 for 1° and 2° RX
Weak Nu^- (H_2O, etc.): S_N1 for 2° and 3° RX
Strong base: E1 and E2

Answers to Problems

5.15   (a) is the only alkyl halide of the group. Compound (b) is an aryl halide, and (c) does not contain a carbon-halogen bond.

5.16   (a) 1-bromo-1-methylcyclohexane   (b) 1,3-dibromobutane

(c) <u>trans</u>-2-chloro-1-cyclopentanol   (d) 3,3,3-trichloro-1-propene

5.17   (a) and (b), $(CH_3)_2CHCH_2I$ (c) ⬡ Cl, Cl   (d) $(CH_3)_2CHCHBrCH_2OH$

5.18   (a) 1°   (b) 1°   (c) methyl   (d) 3°   (e) 3°   (f) 2°   (g) 1°

(h) 2°

5.19   (a) CH_3CH_2Br (b) ⬡—Br   (c) ⬡ CH_3 Br   (d) ⬡ CH_2Br Br

5.20   (a), (c)

5.21   (a) CH_3I because it is a methyl iodide.

(b) ⬡—I because it is an alkyl iodide.

(c) ⬡—Cl because it is an alkyl chloride. (Aryl halides do not undergo this
type of nucleophilic substitution reaction.)

5.22   (a) $CH_3CH_2Br + Na^+ {}^-OCH_2CH_3$ (b) $CH_3CH_2CHClCH_3 + NaOH$

(c) $CH_3CH_2CH_2CH_2Cl + Na^+ {}^-C{\equiv}CH$ (d) $C_6H_5O^- Na^+ + (CH_3)_2CHBr$

(e) $Na^+ {}^-O(CH_2)_4Br$ (f) $CH_3CH_2I + Na^+ {}^-O_2CCH_2CH(CH_3)_2$

(g) $BrCH_2CH_2CH_2Br + 2 Na^+ {}^-C{\equiv}CH$

5.23   (a) ⬡—SCH_3 + NaCl ($^-SCH_3$ is the nucleophile.)

(b) O⬡—$OCH_2CH_2CH_3$ + NaBr (The nucleophile is $CH_3CH_2CH_2O^-$.)

(c) $HOCH_2CH_2CH_2OH$ + NaI (Some $ICH_2CH_2CH_2OH$ might also be observed. The
ratio of the substitution products would depend upon the ratio of reactants.)

5.24   (a), (b), (d)

5.25   (a) CH_3O---C---Br with H_3C H above and CH_2CH_3 below

(b) CH_3O---C---Br with H CH_3 above and CH_2CH_3 below

(c) CH_3O---C---Br with H_3C H above and $CH(CH_3)_2$ below

(d) CH_3O---C---Br with H CH_3 above and $CH(CH_3)_2$ below

The nucleophile must attack the carbon that has the halogen.   The following type of transition state is <u>incorrect</u>:

$$CH_3O\text{-}\text{-}\text{-}CH_3CH_2\overset{\overset{\displaystyle H_3C\quad H}{\diagdown\quad\diagup}}{C}\text{-}\text{-}\text{-}Br$$

<u>incorrect</u>

5.26   (a) (<u>R</u>)- $CH_3\overset{\overset{\displaystyle OH}{|}}{C}H(CH_2)_3CH_3$ (b) (<u>R</u>)-$CH_3CH_2\overset{\overset{\displaystyle CH_3}{|}}{S}CH(CH_2)_3CH_3$

(b) (<u>R</u>)-$CH_3C\equiv C\overset{\overset{\displaystyle CH_3}{|}}{C}H(CH_2)_3CH_3$ (d) (<u>R, R</u>)-$CH_3(CH_2)_3\underset{\underset{\displaystyle CH_2CH_3}{|}}{\overset{\overset{\displaystyle CH_3}{|}}{C}}HOCH(CH_2)_3CH_3$

In each case, the configuration around the chiral carbon atom of (<u>S</u>)-2-iodohexane is inverted.   In (d), the configuration of the attacking nucleophile is not changed in the reaction.

5.27   (a)

Since inversion occurs, only the <u>trans</u> isomer is formed in this case.   The transition state would be:

(b)   HO OH

An intermediate as well as a by-product would be $\underset{Cl}{}$ OH

5.28   (a) $\underline{E}_{act}$ (b) transition-state energy   (c) $\Delta\underline{H}$

5.29   (c)

5.30   rate = $\underline{k}[CH_3I][OH^-]$

(a) The rate is multiplied by 3 × 2, or 6.

(b) The rate is halved.

(c) The rate increases.

(d) The concentration of each reactant is halved; therefore, the rate is divided by four [(1/2) × (1/2) = 1/4].

5.31   (a) $(CH_3CH_2)CHI$, the 2° halide

(b) $(CH_3)_2CHI$, the iodide

(c) ⬡—CH_2Cl, the 1° halide

(d) ⬡—Cl, the 2° halide

5.32   (c), the 3° carbocation.

5.33   (a), (b), (c). The iodide is the fastest; both 3° alkyl halides are faster than the 2°
halide because they yield lower-energy carbocations.

5.34   (a) true   (b) true   (c) false (formation of the carbocation is the rate-determining step)   (d) false (all reactions have transition states)

5.35   (a) $\underline{E}_{act}$ for Step 1   (b) $\underline{E}_{act}$ for Step 2   (c) $\underline{\Delta H}$

5.36   (a) $(CH_3)_3COH$ (b) $\underline{(R)}\underline{(S)}$-$CH_3\overset{\overset{\displaystyle OH}{|}}{C}H(CH_2)_5CH_3$

(c) $\underline{cis}$ and $\underline{trans}$ ⬡—OH
                                    |
                                   CH_3

(d) $(2\underline{R}, 4\underline{S})$ and $(2\underline{S}, 4\underline{S})$-$CH_3\overset{\overset{\displaystyle OH}{|}}{C}HCH_2\overset{\overset{\displaystyle CH_3}{|}}{C}HCH_2CH_3$

In (d), the configuration at position 4 is not changed because this carbon is not involved in the reaction.

5.37   (a) $(CH_3)_2\overset{+}{C}CHCH_2CH_3$ (methyl shift, 2° to 3° carbocation)
                          |
                         CH_3

(b) $CH_2{=}CH\overset{+}{C}HCH_2CH_3$ (hydride shift, 2° to allyl carbocation)

(c) $(CH_3)_2\overset{+}{C}CH_2CH_2CH(CH_3)_2$ (hydride shift, 2° to 3°)

(d) ⬡$^{+}$—CH_2CH_3 (hydride shift, 2° to 3°)

5.38   (a) $\left[(CH_3)_3C\overset{+}{C}HCH_3\right] \longrightarrow \left[(CH_3)_2\overset{+}{C}CH(CH_3)_2\right] \overset{H_2O}{\longrightarrow}$

$$\left[\begin{array}{c} \overset{+}{O}H_2 \\ | \\ (CH_3)_2CCH(CH_3)_2 \end{array}\right] \overset{-H^+}{\rightleftharpoons} \begin{array}{c} OH \\ | \\ (CH_3)_2CCH(CH_3)_2 \end{array}$$

(b) $\left[(CH_3)_2CH\overset{+}{C}HCH_2CH_3\right] \longrightarrow \left[(CH_3)_2\overset{+}{C}CH_2CH_2CH_3\right] \overset{CH_3CH_2OH}{\longrightarrow}$

$$\left[\begin{array}{c} \overset{+}{H}OCH_2CH_3 \\ | \\ (CH_3)_2CCH_2CH_2CH_3 \end{array}\right] \overset{-H^+}{\rightleftharpoons} \begin{array}{c} OCH_2CH_3 \\ | \\ (CH_3)_2CCH_2CH_2CH_3 \end{array}$$

5.39

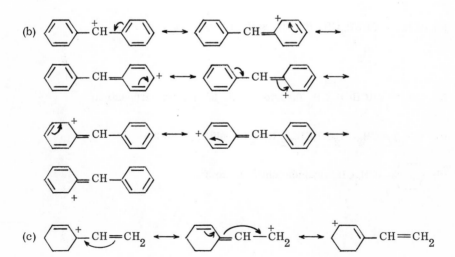

The attacking nucleophile is the product of reaction (a).

5.40   (a)   [naphthalene with CH₂Br and BrCH₂ groups] $\overset{OH^-}{\longrightarrow}$ (b) $CH_2\!\!=\!\!CHCH_2Cl \overset{NaSCH_3}{\longrightarrow}$

5.41   (a) $CH_2\!\!=\!\!CH\!-\!CH\!\!=\!\!CH\!-\!\overset{+}{C}H_2 \longleftrightarrow CH_2\!\!=\!\!CH\!-\!\overset{+}{C}H\!-\!CH\!\!=\!\!CH_2 \longleftrightarrow$

$$\overset{+}{C}H_2\!-\!CH\!\!=\!\!CH\!-\!CH\!\!=\!\!CH_2$$

(b)   [series of resonance structures with benzyl/phenyl cation delocalization]

(c)   [cyclohexadienyl cation resonance structures with CH=CH₂ substituent]

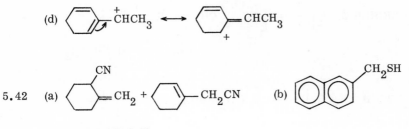

(d) $\overset{+}{C}HCH_3 \longleftrightarrow =CHCH_3$

5.42 (a)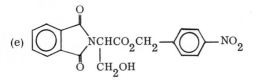

(b) CH_2SH naphthalene

(c) $C_6H_5OCH_2C_6H_5$

(d) $CH_3CH=CHCH=CHCH_2OCH_3 + CH_3CH=CHCHCH=CH_2 +$ with OCH_3 substituent

$CH_3CHCH=CHCH=CH_2$ with OCH_3

(e) phthalimide $NCHCO_2CH_2-\bigcirc-NO_2$ with CH_2OH

5.43 (a) $CH_3CHI(CH_2)_3CH_3 \xrightarrow[(1)]{-I^-} CH_3\overset{+}{C}H(CH_2)_3CH_3 \xrightarrow[(2)]{-H^+} CH_3CH=CH(CH_2)_2CH_3$

(b) Step 1   (c) $CH_2=CH(CH_2)_3CH_3$ (d) Step 2   (e) <u>trans</u>

5.44 (a) 2-butene   (b) 2,3-dimethyl-2-butene   (c) 2-methyl-2-pentene
(d) 1-methyl-1-cyclohexene

In each case, the more substituted alkene is the more stable one.

5.45 (a) $\bigcirc-CH_3$ (b) $\bigcirc-CH_3$ (c) $CH_3-\bigcirc$ (d) $\bigcirc-CH_3$

In each case, elimination gives the more substituted alkene.

5.46 In (a), (c), and (d), there are two <u>trans</u> axial protons (relative to X); elimination
proceeds to the most substituted alkene, just as in E1. In (b), elimination of the
only <u>trans</u> axial proton leads to the least substituted alkene.

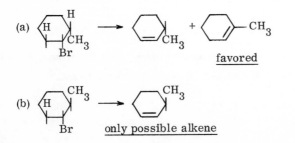

(a) ... $\longrightarrow$ ... $+$ ... $-CH_3$
favored

(b) ... $\longrightarrow$ ...
<u>only possible alkene</u>

5.47   (a) $\left[(CH_3)_2CH\overset{+}{C}HCH(CH_3)_2\right] \longrightarrow \left[(CH_3)_2\overset{+}{C}CH_2CH(CH_3)_2\right] \xrightarrow{-H^+}$

$(CH_3)_2C\!=\!CHCH(CH_3)_2$

(b) $\left[(CH_3)_3C\overset{+}{C}HCH_2CH_2CH_3\right] \longrightarrow \left[\begin{array}{c}(CH_3)_2\overset{+}{C}CHCH_2CH_2CH_3 \\ | \\ CH_3\end{array}\right] \xrightarrow{-H^+}$

$\begin{array}{c}(CH_3)_2C\!=\!CCH_2CH_2CH_3 \\ | \\ CH_3\end{array}$

5.48   (a) $(CH_3)_2CBrCH_2CH_2CH_3$, the 3° halide

(b) $(CH_3)_2CHCHICH_3$, the 2° halide

(c) $CH_3CH_2CH_2Br$, the 1° halide. (The other halide is a vinyl halide and under-goes elimination very slowly, if at all, under the usual conditions.)

(d) CH_3CHICH_3, the iodide

5.49   (a) $CH_3CH_2CHBrCH_3$ (b) $CH_3CH_2CHCH_2Br$ with CH_3 on the carbon   (c)   cyclohexane with CH_2Br and H

(d) $(CH_3CH_2)_2CHI$

5.50   (a) $CH_3CD_2CH_2Br$ (b) cyclohexane ring with CH_3, D, Br (c) $(CH_3)_2CDCH_2I$

(d) $(CD_3)_2CCICD_3$ (e) cyclohexane ring with CH_3, Cl, D, H_3C

Any proton that could be eliminated must be replaced by D.

5.51   (a) and (d), Hofmann   (b) and (c), Saytseff

5.52   (a) $(CH_3)_2CHBr + KOH$ (b) $(CH_3)_3CBr + KOH$

(c) $CH_3CHBrCH_2CH(CH_3)_2 + KOH$ (d) $CH_3CH_2CH_2CH_2CH_2Br + KOH$

In all cases, an alcohol solvent and heat would be used; in (a), (c), and (d), a large base, such as $K^+ \ ^-OC(CH_3)_3$, would minimize substitution reactions.

5.53   (a) E1 or E2: <u>trans</u>-$CH_3CH\!=\!CHCH_2CH_3$

(b) E1 or E2: $(2\underline{E}, 5\underline{E})$-$CH_3CH$=$CHCH_2CH$=$CHCH_3$

(c) E1: $(\underline{E})$-C_6H_5CH=CCH_2CH_3,   E2: $(\underline{Z})$-isomer
$$\underset{\displaystyle C_6H_5}{|}$$

In (c), the E1 product is the more stable $(\underline{E})$-isomer. The E2 product is the $(\underline{Z})$-isomer because H and Br must be <u>anti</u> in the transition state:

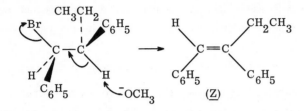

5.54   (a) (1) ⬡—CH_3 (2) ⬡=CH_2

(b) (1) ⬡=$CHCH_3$ (2) ⬡—CH=CH_2

The small base CH_3O^- yields Saytseff products, while the bulky base $(CH_3)_3CO^-$ yields Hofmann products.

5.55   (a), (b), (c), (d)

5.56   (a) S_N2 (b) S_N1 (some S_N2)   (c) E2   (d) S_N1 (some S_N2)

5.57   (a) $(CH_3)_3COCH_3$, S_N1 (b) $C_6H_5OCH_3$, S_N2

(c) $(CH_3)_2C$=$CHCH_3$, E2 and E1   (d) CH_3CH=$CHCH_2CH_3$, E2

(e) CH_2=CCH_2CH_3, E2 and $(CH_3)_2C$=$CHCH_3$, E1
$$\underset{\displaystyle CH_3}{|}$$

5.58   (a) aqueous KOH (a more polar solvent)

(b) 1-bromobutane (a 1° alkyl halide)

(c) warm (high temperature favors elimination)

5.59   (b), because the lower concentration of nucleophile decreases the S_N2 rate but does not affect the S_N1 rate (which leads to rearrangement).

5.60   (a) CH_2=$CHCH_2Cl$ + NaOH   (b) $(CH_3)_2CHCH_2Cl$ + NaSH

(c) $C_6H_5CH_2CHBrC_6H_5$ + KOH in CH_3CH_2OH

(d) [pentane ring with CH_3 at top, H_3C and Cl on adjacent carbons] + $NaOCH_3$

5.61 $CH_3CH_2CHClCH_3$ $\xrightarrow{\text{base}}$ $CH_3CH{=}CHCH_3$ + $CH_3CH_2CH{=}CH_2$

the alkyl halide

All other butyl chlorides would yield only one alkene upon elimination.

5.62   (c), (d), (a), (b). Although (c) is a 3° halide, it cannot form a carbocation because its ring system cannot become planar around the functional carbon atom.

5.63   The reaction proceeds by an S_N2 mechanism and every product molecule is (S). For every one molecule that reacts, a pair of molecules becomes racemic.

$$2\ (\underline{R}){-}RI\ +\ {}^*I^-\ \xrightarrow{S_N2}\ (\underline{R}){-}RI\ +\ (\underline{S}){-}R^*I\ +\ I^-$$

When half the molecules of $(\underline{R})$—RI have undergone reaction, the mixture is racemic. If the reaction had proceeded by an S_N1 mechanism, the rate of racemization would have equalled the rate of $^*I^-$ uptake because both $(\underline{R})$ and $(\underline{S})$ products are obtained.

$$2\ (\underline{R}){-}RI\ +\ 2\ {}^*I^-\ \xrightarrow{S_N1}\ (\underline{R}){-}R^*I\ +\ (\underline{S}){-}R^*I\ +\ 2\ I^-$$

5.64   Although $ClCH_2OCH_3$ is a primary alkyl halide, the carbocation is resonance-stabilized.

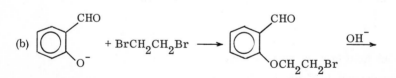

5.65   (a)

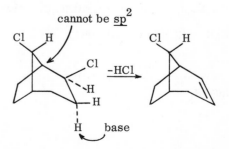

CHO

+ $Br(CH_2)_4Br$ $\xrightarrow{-Br^-}$

(b)

CHO

+ $BrCH_2CH_2Br$ $\longrightarrow$

CHO

OCH_2CH_2Br

$\xrightarrow{OH^-}$

5.66   Elimination yields only one alkene because the geometry of this ring system does not allow pi-bond overlap by a bridgehead carbon (a carbon at the ring juncture). Compound (b) undergoes more-rapid elimination because the base attacks from the less-hindered side of the molecule.

cannot be $\underline{sp}^2$

Cl   H   Cl   H

Cl

$\xrightarrow{-HCl}$

H
H

H   base

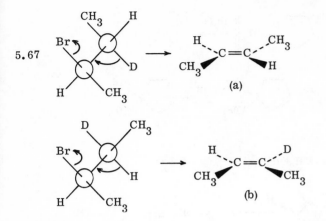

5.67

(a)

(b)

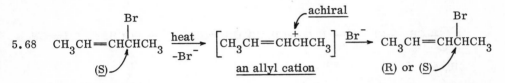

5.68

an allyl cation

5.69   (a) is preferred because (b) would lead to an alkene instead of the desired ether.

5.70   (a) Add aqueous AgNO$_3$.

$$Cl-\!\!\!\bigcirc\!\!\!-CH_3 \xrightarrow[H_2O]{Ag^+} \text{no reaction}$$

$$\bigcirc\!\!\!-CH_2Cl \xrightarrow[H_2O]{Ag^+} \bigcirc\!\!\!-CH_2OH + H^+ + AgCl\!\downarrow$$

(b) Add aqueous AgNO$_3$.

$$\bigcirc\!\!\!-Br \xrightarrow[H_2O]{Ag^+} \text{no reaction}$$

$$\bigcirc\!\!\!-Br \xrightarrow[H_2O]{Ag^+} \bigcirc\!\!\!-OH + H^+ + AgBr\!\downarrow$$

(c) Add KOH in CH$_3$CH$_2$OH and heat. Compare the physical constants of the product with those of known alkenes.

(2$\underline{S}$, 3$\underline{R}$)-isomer $\longrightarrow$ ($\underline{Z}$)-3-methyl-2-pentene

(2$\underline{S}$, 3$\underline{S}$)-isomer $\longrightarrow$ ($\underline{E}$)-3-methyl-2-pentene

5.71   (a) $(CH_3)_2CHCl \xrightarrow[-AgCl]{Ag^+} \left[(CH_3)_2\overset{+}{C}H\right] \xrightarrow{H_2O} (CH_3)_2CH\overset{+}{O}H_2 \underset{}{\overset{-H^+}{\rightleftarrows}} (CH_3)_2CHOH$

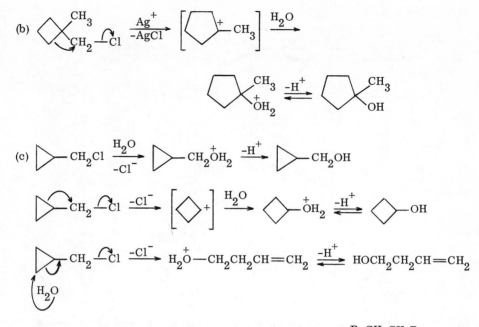

$$5.72 \quad BrCH_2CH_2Br \xrightarrow[-2\ Br^-]{OH^-} HOCH_2CH_2OH \xrightarrow{Na} {}^-OCH_2CH_2O^- \xrightarrow[-2\ Br^-]{BrCH_2CH_2Br}$$

Chapter 6

Free-Radical Reactions;
Organometallic Compounds

Some Important Features

A free radical is an atom or group of atoms containing an unpaired electron; most free radicals are highly reactive intermediates. An alkyl free radical is achiral around the reacting carbon.

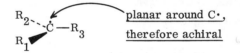

planar around C·,
therefore achiral

Free-radical reactions occur stepwise: (1) initiation (initial generation of free radicals); (2) propagation (reaction of free radicals yielding new reactive free radicals); (3) termination (destruction of free radicals, often by the joining together of two free radicals or by the formation of a relatively stable, nonreactive free radical).

Different hydrogen atoms in a structure may be replaced by halogen atoms at different rates. If the intermediate free radical is stabilized, then reaction rate at that position is enhanced.

$$CH_4 \quad RCH_3 \quad R_2CH_2 \quad R_3CH \quad \text{allylic and benzylic}$$

Increasing rate of free-radical reaction and increasing
stability of free radical when H· is removed

Because of the enhanced reactivity of some H atoms, we can often predict the principal free-radical product.

$$\overset{1°}{\underset{}{\searrow}} \overset{2°}{\underset{}{\searrow}} \quad CH_3CH_2CH_3 \xrightarrow[-H·]{X_2, h\nu} [CH_3\dot{C}HCH_3 \text{ favored over } CH_3CH_2\dot{C}H_2] \xrightarrow{-X·} CH_3\overset{X}{\underset{|}{C}HCH_3}$$

The reagent also affects the product ratio. For example, a less reactive free-radical reagent is more selective.

$$\text{CH}_3\text{CH}_2\text{CH}_3 \begin{cases} \xrightarrow[\text{(more reactive)}]{\text{Cl}_2, h\nu} \text{CH}_3\text{CHClCH}_3 + \text{CH}_3\text{CH}_2\text{CH}_2\text{Cl} \\ \\ \xrightarrow[\text{(less reactive)}]{\text{Br}_2, h\nu} \text{CH}_3\text{CHBrCH}_3 \end{cases}$$

The mechanisms of free-radical reactions are discussed in the text, including the reasons for the reactivities of different H atoms in a structure, the kinetic isotope effect, the different reactivities of halogenating agents, and halogenation with NBS. Other types of free-radical reactions (pyrolysis, auto-oxidation, and the quinone-hydroquinone reaction) are also covered in this chapter. Free-radical initiators and inhibitors are mentioned.

Organometallic compounds are compounds containing carbon-metal bonds. One of the most important organometallic compounds is the Grignard reagent, RMgX.

The Grignard reagent contains a carbanion-like organic group that reacts with a compound containing a partially positive hydrogen (as in H_2O or ROH) or a partially positive carbon (as in a carbonyl compound).

$$R\!-\!MgX + \text{H}_2\text{O} \longrightarrow RH + HO\!-\!MgX$$

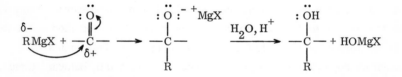

Lithium reagents (RLi) behave like Grignard reagents. Cuprates (R_2CuLi) are used to prepare alkanes.

$$2\,RLi \xrightarrow{\text{CuI}} \text{R}_2\text{CuLi} \xrightarrow{R'X} R\!-\!R'$$

Reminders

Be sure you know how to write resonance structures for free-radical intermediates. In these cases, we shift only one electron (not two, as in carbocation intermediates).

$$\underset{\text{allylic H}}{R_2\text{C}\!=\!\text{CHCH}_2\text{R}} \xrightarrow[\text{-HX}]{X\cdot} \left[R_2\text{C}\!\overset{\frown}{=}\!\text{CH}\!-\!\dot{\text{C}}\text{HR} \longleftrightarrow R_2\dot{\text{C}}\!-\!\text{CH}\!=\!\text{CHR} \right]$$

an allyl free radical

a resonance-stabilized intermediate

$$\text{RÖCH}_2\text{R} \xrightarrow[\text{-HX}]{\text{X·}} \left[\text{RÖ}\overset{\curvearrowright}{\text{—}}\text{ĊHR} \longleftrightarrow \text{RÖ}\text{=CHR} \right]$$

<u>an ether</u>   <u>a resonance-stabilized intermediate</u>

Remember that a stabilized organic <u>intermediate</u> (relative to reactants) means a faster reaction. (Allylic hydrogens are easily removed by free radicals because the allylic free radical is resonance-stabilized.) Conversely, a stabilized <u>reactant</u> (relative to intermediates) means a slower reaction. (I· is more stable and thus less reactive than Cl·.)

$$\text{X· + RH} \longrightarrow \text{R· + HX}$$

<u>stabilization means</u>   <u>stabilization means</u>
<u>a slower reaction</u>   <u>a faster reaction</u>

A Grignard reagent is a very strong base that reacts with acidic hydrogens. Acid-base reactions, such as $\text{RMgX} + \text{H}_2\text{O} \longrightarrow \text{RH} + \text{HOMgX}$, are fast compared to most organic reactions, such as that of RMgX with a carbonyl compound.

Answers to Problems

6.20   (a)
$$
\begin{array}{cc}
\text{H} & \text{H} \\
\text{H} : \text{C} : \text{C} : \ddot{\text{O}} \cdot \\
\text{H} & \text{H}
\end{array}
$$
(b)
$$
\begin{array}{c}
\text{H} \cdot \\
\quad \ddot{\text{C}} :: \text{C} : \text{C} : \text{H} \\
\text{H} \cdot \quad\quad \text{H} \quad \text{H}
\end{array}
$$

6.21   (a) and (e) are propagation steps   (b) is initiation
       (c) and (d) are termination

These different steps can be distinguished as follows: If a free radical is formed from a nonradical reactant, the reaction is an initiation step. If both a reactant and a product are free radicals, the reaction is a propagation step. If free-radical reactants lead to nonradical products, the reaction is a termination step.

6.22 $\text{Cl}_2 \xrightarrow{h\nu} 2 \text{ Cl·}$

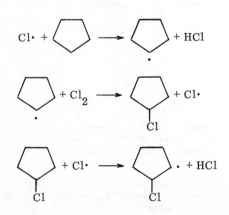

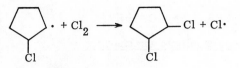

6.23   The ratio would be the same as the ratio of the types of H being extracted: two propane CH_2 hydrogens, six propane CH_3 hydrogens, twelve cyclohexane CH_2 hydrogens, or 1 : 3 : 6. The product ratio would therefore be as follows:

1 part $CH_3CHClCH_3$ 3 parts $CH_3CH_2CH_2Cl$ 6 parts ⟨ ⟩—Cl

6.24   (a) $ClCH_2CHClCH_2Cl$, (R)-$Cl_2CHC\overset{*}{H}ClCH_3$, $ClCH_2CCl_2CH_3$

(b) $ClCH_2CHClCH_2Cl$, $Cl_2CHCHClCH_3$, $ClCH_2CCl_2CH_3$

(c) (S)-$ClCH_2\overset{*}{C}HClCH_2CH_3$, $CH_3CCl_2CH_2CH_3$,

(R)-$CH_3\overset{*}{C}HClCHClCH_3$, (R)-$CH_3\overset{*}{C}HClCH_2CH_2Cl$

In each case, chlorination at a chiral carbon yields a racemic or achiral product, while chlorination at an achiral carbon does not affect the stereochemistry of other, chiral carbons. In (a), note that chirality can be lost by the formation of identical groups on a chiral carbon.

6.25 $(CH_3)_4C$

6.26   (d), (b), (a), (c), (e)

6.27   (b), (a), (c), ranked in the order of stability of the intermediate free radical.

6.28   (a) [structure] ⟨ ⟩=CH_2 ⟷ ⟨ ⟩—ĊH_2

(b) [structure] ⟨ ⟩—ĊHCH_3 ⟷ ⟨ ⟩—CHCH_3 ⟷

⟨ ⟩=CHCH_3 ⟷ ⟨ ⟩=CHCH_3

6.29   (a) [structure] ⟨ ⟩—ĊO: ⟷ ⟨ ⟩—Ċ=O:

(b) [structure] ⟨ ⟩—Ö: ⟷ ⟨ ⟩=O ⟷ ⟨ ⟩=O ⟷ ⟨ ⟩=O

(c) [structure] ⟨ ⟩—ṄCH_3 ⟷ ⟨ ⟩=NCH_3 ⟷ ⟨ ⟩=NCH_3 ⟷ ⟨ ⟩=NCH_3

6.30   (b), (a), (c), (d)

6.31   (a) 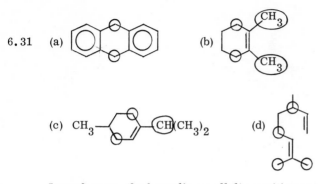   (b)

(c) CH₃— ⟨ring⟩—CH(CH₃)₂   (d)

In each case, the benzylic or allylic positions are attacked.

6.32   (a)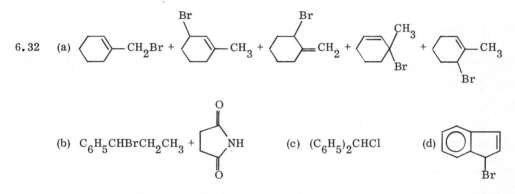

(b) C₆H₅CHBrCH₂CH₃ + ⟨imide⟩NH   (c) (C₆H₅)₂CHCl   (d) ⟨indene⟩Br

In (a), three different free radicals can be formed.   Each is resonance-stabilized; therefore, bromination can occur at five positions.

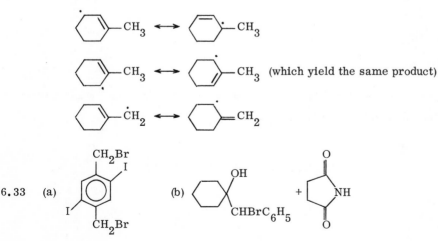

6.33   (a)   (b)   + ⟨imide⟩NH

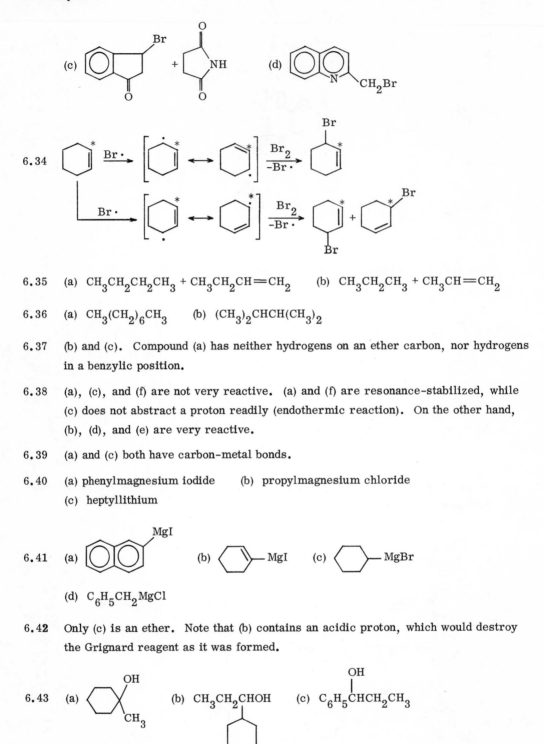

(c)   [structure: Br-substituted indanone] + [succinimide-like structure with NH]   (d)   [quinoline structure with CH₂Br]

6.34   [reaction scheme with Br· and Br₂ steps]

6.35   (a) $CH_3CH_2CH_2CH_3 + CH_3CH_2CH=CH_2$ (b) $CH_3CH_2CH_3 + CH_3CH=CH_2$

6.36   (a) $CH_3(CH_2)_6CH_3$ (b) $(CH_3)_2CHCH(CH_3)_2$

6.37   (b) and (c). Compound (a) has neither hydrogens on an ether carbon, nor hydrogens in a benzylic position.

6.38   (a), (c), and (f) are not very reactive. (a) and (f) are resonance-stabilized, while (c) does not abstract a proton readily (endothermic reaction). On the other hand, (b), (d), and (e) are very reactive.

6.39   (a) and (c) both have carbon-metal bonds.

6.40   (a) phenylmagnesium iodide   (b) propylmagnesium chloride
       (c) heptyllithium

6.41   (a)   [naphthalene with MgI]   (b)   [cyclohexenyl—MgI]   (c)   [cyclohexyl—MgBr]

       (d) $C_6H_5CH_2MgCl$

6.42   Only (c) is an ether. Note that (b) contains an acidic proton, which would destroy the Grignard reagent as it was formed.

6.43   (a)   [cyclohexane with OH and CH₃]   (b) CH_3CH_2CHOH [with cyclohexyl group]   (c) $C_6H_5CHCH_2CH_3$ [with OH]

       (d)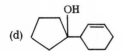

6.44 (a) $(CH_3)_2CHBr \xrightarrow[\text{ether}]{Mg} (CH_3)_2CHMgBr \xrightarrow[\text{(2) } H_2O, H^+]{\text{(1) } (CH_3)_2C=O} (CH_3)_2CH\overset{\displaystyle OH}{\underset{\displaystyle |}{C}}(CH_3)_2$

(b) $(CH_3)_2CHMgBr \xrightarrow[\text{(2) } H_2O, H^+]{\text{(1) } CH_3CHO} (CH_3)_2CH\overset{\displaystyle OH}{\underset{\displaystyle |}{C}}HCH_3$

6.45 (b) $\overset{\frown}{(H}OCH_2CH_2O\overset{\frown}{H)}$ (c) $\overset{\frown}{(H_2}NCH_2CH_2CH_2O_2CCH_3$ (d) $CH(CO_2\overset{\frown}{H})_3$

6.46 (a) ⬡—Li + LiBr   (b) $CH_3CH_2OH + NaBr$

(c) $C_6H_5CH_2CH_2CO_2H + Li^+ + H_2O$ (d) ⬠ + ⬡N⁻ Li⁺

(e) ⬡ + ⬡—$\overset{\displaystyle O}{\overset{\displaystyle ||}{C}}H$
O⁻ ⁺MgBr

The reaction of an acidic proton with a Grignard reagent is much faster than the reaction of a carbonyl group; therefore, only the acid-base reaction products are formed in (e).

6.47 (a) $CH_3CH_2Br \xrightarrow[\text{(2) } CuI]{\text{(1) } Li} (CH_3CH_2)_2CuLi \xrightarrow{CH_3CH_2CH_2Br}$

(b) $CH_3(CH_2)_3Br \xrightarrow[\text{(2) } CuI]{\text{(1) } Li} [CH_3(CH_2)_3]_2CuLi \xrightarrow{CH_3CH_2CH_2CH_2Br}$

(c) $(CH_3)_2CHBr \xrightarrow[\text{(2) } CuI]{\text{(1) } Li} [(CH_3)_2CH]_2CuLi \xrightarrow{CH_3CH_2CH_2CH_2Br}$

6.48 If HBr is present in the reaction mixture, an exchange reaction leading to $C_6H_5CH_3$ occurs. The presence of $C_6H_5CH_3$ leads to a decreased isotope effect.

$$C_6H_5\overset{\displaystyle \cdot}{C}H_2 + HBr \longrightarrow C_6H_5CH_3 + Br\cdot$$

<u>no isotope effect</u>

6.49 (a) suffers from greater steric hindrance than (b) and is therefore of higher energy than (b). Some of the steric hindrance is relieved in the free radical. (See Section 5.7D.)

6.50 The free-radical intermediate in the formation of the allyl Grignard reagent is resonance-stabilized; therefore, <u>two</u> Grignard reagents (and two Grignard products) can be formed.

$$CH_3CH{=}CHCH_2Cl \xrightarrow{\text{Mg}} \left[CH_3CH{=}CH\overset{\cdot}{C}H_2 \longleftrightarrow CH_3\overset{\cdot}{C}HCH{=}CH_2 + M\overset{\cdot}{g}Cl \right]$$

$$\longrightarrow CH_3CH{=}CHCH_2MgCl + CH_3\overset{\overset{\displaystyle MgCl}{|}}{C}HCH{=}CH_2$$

$$\xrightarrow[\text{(2) } H_2O,\ H^+]{\text{(1) } CH_3\overset{\overset{\displaystyle O}{\|}}{C}CH_3} CH_3CH{=}CHCH_2\overset{\overset{\displaystyle CH_3}{|}}{\underset{\underset{\displaystyle CH_3}{|}}{C}}OH$$

$$\text{and } CH_2{=}CHCH{-}\overset{\overset{\displaystyle CH_3}{|}}{\underset{\underset{\displaystyle CH_3}{|}}{C}}OH$$
$$\underset{\displaystyle CH_3}{\overset{|}{}}$$

6.51   (a) $(C_6H_5)_2CHBr \xrightarrow{\text{Mg}} (C_6H_5)_2CHMgBr \xrightarrow{D_2O} (C_6H_5)_2CHD$

(b) $(CH_3)_2CHBr \xrightarrow{\text{Mg}} (CH_3)_2CHMgBr \xrightarrow{D_2O} (CH_3)_2CHD$

(c) $CH_2{=}CHCH_2Br \xrightarrow{\text{Mg}} CH_2{=}CHCH_2MgBr \xrightarrow{D_2O} CH_2{=}CHCH_2D$

6.52   (a) $(C_6H_5)_2CH_2 \xrightarrow{\text{NBS}}$ (b) $(CH_3)_2CH_2 \xrightarrow[\text{h}\nu]{Br_2}$ (c) $CH_2{=}CHCH_3 \xrightarrow{\text{NBS}}$

6.53

inverted, but still (R)
because of a change in
priorities

6.54   This type of problem is called a road map problem and is typical of what a chemist
encounters in structure determination. To approach a road map problem, first
write out a flow diagram:

$$CH_3CH_2CH_2CH_2CH_3 \xrightarrow[h\nu]{Br_2} A + B \xrightarrow[E2]{CH_3O^-} C$$

n-pentane   two bromopentanes   a pentene

We have added what is immediately apparent about the products of the two reactions: A and B are bromopentanes, while C must be an elimination product (and therefore a pentene). All we need do is fill in the positions of substitution. n-Pentane is most likely to yield 2- and 3-bromopentane (A and B). Will these two compounds yield the same alkene (C)? The answer is yes. Now, we fill in the flow diagram with structures.

$$CH_3CH_2CH_2CH_2CH_3 \xrightarrow[h\nu]{Br_2} CH_3CHBrCH_2CH_2CH_3 + CH_3CH_2CHBrCH_2CH_3$$

                                    A   B

$$\xrightarrow{E2} CH_3CH = CHCH_2CH_3$$

                         C

Chapter 7

Alcohols, Ethers, and Related Compounds

Some Important Features

Because the O in R$\overset{\cdot\cdot}{\text{O}}$H or R$\overset{\cdot\cdot}{\text{O}}$R is electronegative and has unshared electrons, alcohols, phenols, and ethers undergo hydrogen bonding with compounds that contain partially

positive hydrogens $\left(\begin{array}{c} \text{H} \\ | \\ \text{R}\overset{\cdot\cdot}{\text{O}}\text{H---}\overset{\cdot\cdot}{\text{O}}\text{R or R}_2\text{O---H}_2\text{O} \end{array} \right)$. Alcohols and ethers are protonated in

acidic solution. Alcohols and, to a limited extent, ethers undergo substitution reactions with HX (3° ROH > 2° ROH > 1° ROH). Because 3° and 2° alcohols react by an S_N1 path with HX, PCl$_3$ and SOCl$_2$ are often used as reagents for converting alcohols to alkyl halides. Both PCl$_3$ and SOCl$_2$ give stereospecific reactions without rearrangement.

Dehydration of alcohols yields Saytseff products. (Again, 3° ROH > 2° ROH > 1° ROH.)

Alcohols act as acids when treated with extremely strong bases (such as RMgX) or with alkali metals (Na or K). Phenols are more acidic than alcohols because the phenoxide ion is resonance-stabilized; phenols react with NaOH.

Other important topics covered in this chapter are the reactions of epoxides in acidic and basic solutions; inorganic esters of alcohols; oxidation of 1° and 2° alcohols; and thiols and their derivatives.

Reminders

In acidic solution, an alcohol is protonated and can be attacked by a nucleophile such as Br$^-$.

In dilute base, alcohols are not protonated and undergo no appreciable reaction.

Reaction of a 3° or 2° alcohol in acidic solution yields <u>racemic</u> or <u>achiral</u> products.

Answers to Problems

7.32   (a) ether   (b) phenol (a hydroquinone)   (c), (d) alcohols.

(d) is not a phenol because the OH group is not attached to a ring carbon; rather, the compound is a benzyl alcohol.

7.33 (a) $(CH_3)_2NCH_2CH_2OCH(C_6H_5)_2$
$\quad$ ether

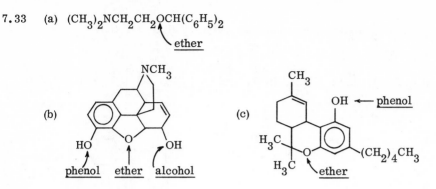

(b) phenol $\quad$ ether $\quad$ alcohol

(c) CH$_3$ $\quad$ OH ← phenol $\quad$ ether

7.34 (a) <u>t</u>-butyl alcohol, because its alkyl group is the most branched and thus the least hydrophobic.

(b) tetrahydrofuran, because of its compactness and because its oxygen is more exposed.

(c) 1-octanol, because of the hydrophilic OH group.

(d) 1,5-pentanediol, because it has <u>two</u> OH groups.

7.35 (a) isopropyl alcohol or 2-propanol, 2° $\quad$ (b) 3-methyl-1-butanol, 1°

(c) <u>sec</u>-butyl alcohol or 2-butanol, 2° $\quad$ (d) 2,5-heptanediol, both 2°

(e) 4-phenyl-2-pentanol, 2° $\quad$ (f) <u>trans</u>-3-methyl-1-cyclohexanol, 2°

(g) <u>cis</u>-2-methyl-1-cycloheptanol, 2° $\quad$ (h) <u>cis</u>-3-penten-1-ol, 1° (but not allylic).

7.36 (a) 1,2-dimethoxyethane $\quad$ (b) isopropyl propyl ether, 2-propoxypropane, or 1-isopropoxypropane $\quad$ (c) <u>cis</u>-2,3-dimethyloxirane $\quad$ (d) <u>trans</u>-2,3-dimethyloxirane.

7.37 (a) $(CH_3)_2CHCH_2CH_2OH$ $\quad$ (b) $CH_2{=}C(CH_3)_2$ (see Chapter 5)

(c) CH$_2$OH $\quad$ (d) OH $|$ CHCH(CH$_3$)$_2$ $\quad$ (e) CH$_3$ OH

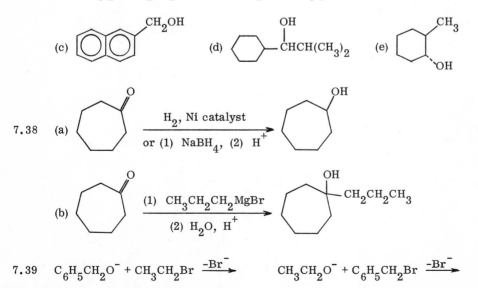

7.38 (a) $\xrightarrow[\text{or (1) NaBH}_4\text{, (2) H}^+]{\text{H}_2\text{, Ni catalyst}}$

(b) $\xrightarrow[\text{(2) H}_2\text{O, H}^+]{\text{(1) CH}_3\text{CH}_2\text{CH}_2\text{MgBr}}$ CH$_2$CH$_2$CH$_3$

7.39 $C_6H_5CH_2O^- + CH_3CH_2Br \xrightarrow{-Br^-}$ $\quad\quad$ $CH_3CH_2O^- + C_6H_5CH_2Br \xrightarrow{-Br^-}$

7.40 $C_6H_5O^- + CH_3CH_2CH_2Br \xrightarrow{-Br^-}$ but not $CH_3CH_2CH_2O^- + C_6H_5X$

7.41   (a) $(CH_3)_2CHOH \xrightarrow{H^+} \rightleftharpoons (CH_3)_2CHOH_2^+ \xrightarrow{-H_2O} \rightleftharpoons \left[(CH_3)_2CH^+\right] \xrightarrow{I^-} (CH_3)_2CHI$

   2° alcohol, S_N1

(b)   ⬡—OH $\xrightarrow{H^+} \rightleftharpoons$ ⬡—$\overset{+}{O}H_2$ $\xrightarrow{-H_2O} \rightleftharpoons$ $\left[⬡ +\right] \xrightarrow{I^-}$ ⬡—I

(c) $CH_3CH_2CH_2CH_2OH \xrightarrow{H^+} \rightleftharpoons CH_3CH_2CH_2CH_2\overset{+}{O}H_2 \xrightarrow{I^-}$

   1° alcohol, S_N2

$$\left[CH_3CH_2CH_2\overset{\overset{\delta+}{OH_2}}{\underset{\underset{\delta-}{I}}{\overset{|}{\underset{|}{C}}H_2}} \right] \xrightarrow{-H_2O} CH_3CH_2CH_2CH_2I$$

7.42   (a) $(CH_3)_2CHCH_2CH_2Br + H_2O$ (b) ⬡—Cl + H_2O

(c) $C_6H_5CH_2CH_2I + H_2O$ (d)

[structure with CH₃ and Br and CH(CH₃)₂]   +   [structure with CH₃ and Br and CH(CH₃)₂]

In (c), we would also expect some $C_6H_5CH{=\!=}CH_2$. In (d), both the cis and the trans product are obtained because the reaction proceeds by way of a carbocation.

7.43   (a) [naphthalene with CH₂Br]   + H_2O (fastest because it is a benzyl position)

(b) $CH_3CH_2CH_2Br + H_2O$ (slowest because it is a 1° position)

(c) [cyclohexane with CH₃ and Br]   + H_2O

7.44   (b) $CH_3CH_2CH_2OH \xrightarrow{H^+} \rightleftharpoons CH_3CH_2CH_2\overset{+}{O}H_2 \xrightarrow{Br^-}$

   1° alcohol, S_N2

$$\left[\underset{\underset{CH_2CH_3}{|}}{\overset{\overset{H \quad H}{\diagdown \diagup}}{\underset{\delta-}{Br}\text{---}C\text{---}\overset{\delta+}{OH_2}}} \right] \xrightarrow{-H_2O} CH_3CH_2CH_2Br$$

(c)

$$\text{(cyclohexane with CH}_3\text{ and OH)} \xrightarrow{H^+} \text{(cyclohexane with CH}_3\text{ and } \overset{+}{O}H_2) \xrightarrow{-H_2O} \left[\text{(cyclohexyl}^+ \text{—CH}_3)\right] \xrightarrow{Br^-} \text{(cyclohexane with CH}_3\text{ and Br)}$$

2° alcohol, S_N1

7.45

$$\left[\begin{array}{c} R\;H \\ \delta-\;\;\overset{\backslash\;/}{}\;\;\delta+ \\ Br---C---\overset{+}{O}H_2 \\ | \\ H \end{array}\right]$$

E ↑

(energy diagram, single barrier)

$\overset{+}{R}OH_2 + Br^-$ $RBr + H_2O$

Progress of (b) →

$[\overset{+}{R}] + H_2O + Br^-$

E ↑

(energy diagram, double barrier)

$\overset{+}{R}OH_2 + Br^-$ $RBr + H_2O$

Progress of (c) →

7.46 (a)

$$(CH_3)_3C\overset{\overset{\displaystyle OH}{|}}{C}HCH_3 \xrightarrow[\;\;]{H^+} \xrightarrow{-H_2O} \left[(CH_3)_3C\overset{+}{C}HCH_3\right] \xrightarrow{\text{methyl shift}}$$

a 2° carbocation

$$\left[(CH_3)_2\overset{+}{C}\overset{\overset{\displaystyle CH_3}{|}}{C}HCH_3\right] \xrightarrow{Cl^-} (CH_3)_2CClCH(CH_3)_2$$

a 3° carbocation

(b)

$$(C_6H_5)_2CHCH_2OH \xrightarrow{H^+} (C_6H_5)_2CHCH_2\overset{+}{O}H_2 \xrightarrow[-H_2O]{\text{hydride shift}}$$

$$\left[(C_6H_5)_2\overset{+}{C}CH_3\right] \xrightarrow{I^-} (C_6H_5)_2CICH_3$$

resonance-stabilized
by two rings

7.47 (a) $(CH_3)_2C=C(CH_3)_2$, Saytseff product from a 3° carbocation.

(b) $(C_6H_5)_2C=CH_2$, only possible alkene from the rearranged carbocation.

7.48 (a) HCl undergoes reaction with t-butyl alcohol, not with n-butyl alcohol.

(b) Hot HI undergoes reaction with diethyl ether, not with pentane.

7.49 (a) (R)-CH$_3$CHCl(CH$_2$)$_3$CH$_3$ (inverted)   (b) (S)-CH$_3$CHCl(CH$_2$)$_3$CH$_3$ (retention)

(c) CH$_3$CHCl(CH$_2$)$_3$CH$_3$ (racemic from S_N1 reaction)

7.50 (a) trans-CH$_3$CH=CHCH$_2$CH$_2$CH$_3$ (b) trans-C$_6$H$_5$CH=CHCH$_3$

(c) $CH_2=CHCH_2CH_3$ (d) trans-$CH_3CH=CHCH_3$

(e) $HOCH_2CH_2CH=C(CH_3)_2$

In each case, the more substituted alkene (trans, if possible) is formed. In (e), the 4-OH is eliminated preferentially because 3° alcohols undergo elimination more readily than 1° alcohols.

7.51

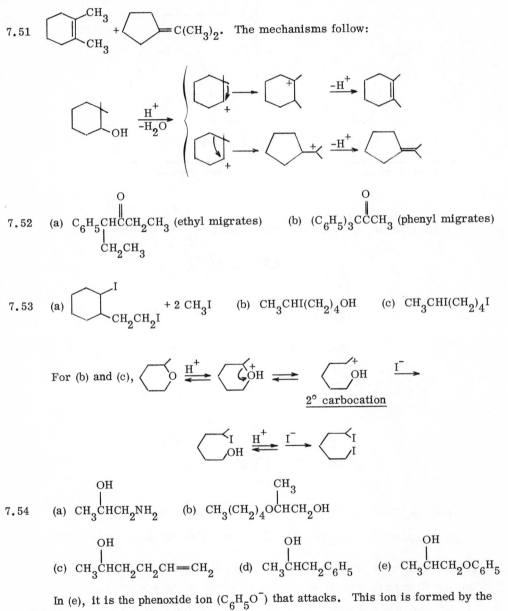

7.52   (a) $C_6H_5CHCCH_2CH_3$ (ethyl migrates)   (b) $(C_6H_5)_3CCCH_3$ (phenyl migrates)

7.53   (a) ... + 2 CH_3I (b) $CH_3CHI(CH_2)_4OH$ (c) $CH_3CHI(CH_2)_4I$

For (b) and (c), ...   2° carbocation

             OH CH_3
7.54   (a) $CH_3CHCH_2NH_2$ (b) $CH_3(CH_2)_4OCHCH_2OH$

             OH   OH   OH
       (c) $CH_3CHCH_2CH_2CH=CH_2$ (d) $CH_3CHCH_2C_6H_5$ (e) $CH_3CHCH_2OC_6H_5$

In (e), it is the phenoxide ion ($C_6H_5O^-$) that attacks. This ion is formed by the reaction of phenol and NaOH.

7.55   The attack on the unsymmetrical epoxide is by an acidic reagent; consequently, Br$^-$ attacks at the 3° carbon. Since the bromide attacks from the opposite side of the bridging three-membered ring, a <u>trans</u>-bromohydrin is the product.

7.56   (a) $CH_3CH_2\overset{+}{O}H_2$ (b) 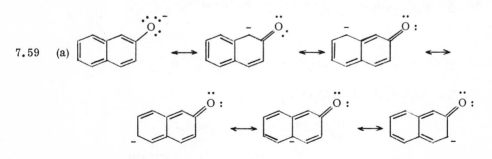 $-O^- + H_2O$ (c) $+ H_2O$

(d) $(CH_3)_3COH + OH^-$

In (b), "no appreciable reaction" is also an acceptable answer because alcohols are weaker acids than water and very little alkoxide would be formed.

7.57   (a) $(CH_3)_2CHOCH_2CH_3 + CH_2{=}CHCH_3$, primarily by S_N2 and E2 reactions of an alkyl halide with a strong nucleophile and strong base.

(b) no reaction between an ether and a base.

(c) $NaOH + CH_3CH_2OH$, acid-base reaction to yield a weaker acid (CH_3CH_2OH) and a weaker base $(NaOH)$.

(d) $CH_3CO_2^- + CH_3CH_2OH$, acid-base reaction.

(e) $C_6H_5O^- Na^+ + CH_3CH_2OH$, acid-base reaction.

7.58   (a) $CH_3CH_2CH_2CH_2O^- \overset{+}{M}gI + CH_4$ (b) $CH_3CH_2CH_2CH_2O^- Li^+ + C_6H_6$

(c) no appreciable reaction   (d) no reaction   (e) $CH_3(CH_2)_3Br + H_2O$

(f) $CH_3CH_2CH_2CH_2O^- K^+ + H_2$

The reagents in (a) and (b) are both strong bases and remove a proton from 1-butanol. The reagents in (c) and (d) are not strong enough bases to react.

7.59   (a)

(b) The alkoxide ion does not have the negative charge in allylic position; there is no resonance-stabilization.

$$CH_2{=}CH-CH_2-\ddot{\underset{\cdot\cdot}{O}}:^-$$

(c) no resonance-stabilization.

7.60   (a) $(CH_3)_2CHOH \xrightarrow{Na} (CH_3)_2CHO^- \xrightarrow{\underset{O}{\overset{\triangle}{CH_2CHCH_3}}} (CH_3)_2CHOCH_2\overset{\underset{\displaystyle OH}{|}}{C}HCH_3$

(b) $CH_3CH_2OH \xrightarrow{HBr} CH_3CH_2Br \xrightarrow{(CH_3)_2CHO^-} (CH_3)_2CHOCH_2CH_3$

7.61 $\underset{HO}{\diagup}\diagdown\underset{Cl}{\diagup} \xrightarrow{\text{Na or NaNH}_2} {}^-O\diagdown\curvearrowright Cl \xrightarrow{S_N2} \bigtriangleup_O + Cl^-$

7.62   (a) achiral $CH_3(CH_2)_4\overset{\displaystyle O}{\overset{\|}{C}}CH_3$ (b) racemic $CH_3(CH_2)_4CHICH_3$ (by an S_N1 reac-

tion)   (c) $(\underline{R})-CH_3(CH_2)_4\overset{\underset{\displaystyle O^-}{|}}{C}HCH_3$ Li^+ (chiral carbon unaffected)

(d) achiral $\underline{trans}-CH_3(CH_2)_3CH{=}CHCH_3$ (e) $(\underline{R})-CH_3(CH_2)_4\overset{\underset{\displaystyle O^-}{|}}{C}HCH_3$ $\overset{+}{M}gI + CH_4$

(chiral carbon unaffected)   (f) no reaction   (g) no appreciable reaction

(h) $(\underline{R})-CH_3(CH_2)_4CHClCH_3$

7.63   (a) $(\underline{S})-CH_3\overset{\underset{\displaystyle OH}{|}}{C}H(CH_2)_2CH_3 + TsCl \xrightarrow{-HCl} (\underline{S})-CH_3\overset{\underset{\displaystyle OTs}{|}}{C}H(CH_2)_2CH_3$

(b) $(\underline{R})-CH_3CH_2\overset{\underset{\displaystyle OH}{|}}{C}HCH_3 \xrightarrow[\text{heat}]{H_2SO_4} \underline{trans}-CH_3CH{=}CHCH_3 + H_2O$

(c) $(\underline{R})-CH_3CH_2\overset{\underset{\displaystyle OH}{|}}{C}HCH_3 + ClSO_3H \longrightarrow (\underline{R})-CH_3CH_2\overset{\underset{\displaystyle OSO_3H}{|}}{C}HCH_3 + HCl$

(d) $C_6H_5CO_2H + CH_3CH_2OSO_2OCH_2CH_3 \longrightarrow$

$C_6H_5CO_2CH_2CH_3 + HOSO_2OCH_2CH_3$

(e) $(\underline{R})-CH_3\overset{\underset{\displaystyle OTs}{|}}{C}HCH_2CH_3 + CH_3CH_2OH \longrightarrow (\underline{R})(\underline{S})-CH_3\overset{\underset{\displaystyle OCH_2CH_3}{|}}{C}HCH_2CH_3$

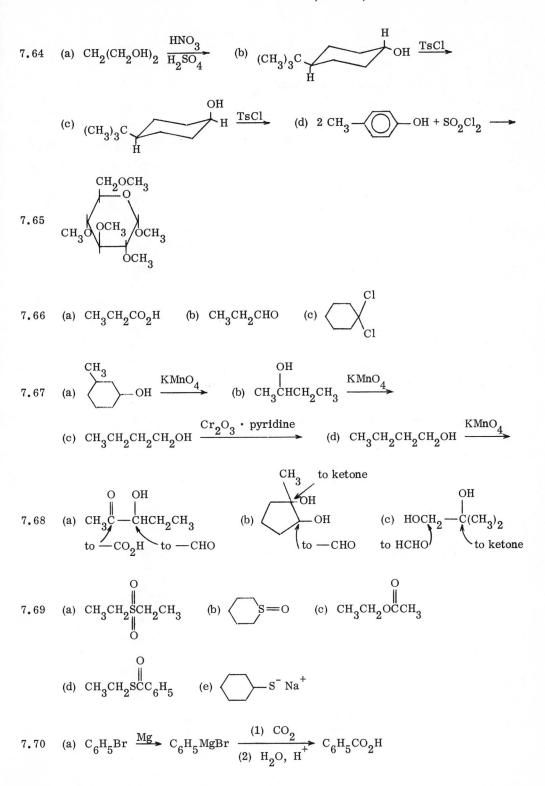

7.64   (a) $CH_2(CH_2OH)_2 \xrightarrow[H_2SO_4]{HNO_3}$ (b) (diagram)

(c) (diagram) $\xrightarrow{TsCl}$ (d) $2\ CH_3-\!\!\langle\bigcirc\rangle\!\!-OH + SO_2Cl_2 \longrightarrow$

7.65   (diagram)

7.66   (a) $CH_3CH_2CO_2H$ (b) CH_3CH_2CHO (c) (diagram)

7.67   (a) (diagram) $-OH \xrightarrow{KMnO_4}$ (b) $CH_3\overset{OH}{\underset{|}{C}}HCH_2CH_3 \xrightarrow{KMnO_4}$

(c) $CH_3CH_2CH_2CH_2OH \xrightarrow{Cr_2O_3\ \cdot\ pyridine}$ (d) $CH_3CH_2CH_2CH_2OH \xrightarrow{KMnO_4}$

7.68   (a) $CH_3\overset{O}{\overset{||}{C}}-\overset{OH}{\underset{|}{C}}HCH_2CH_3$ (b) (diagram)   (c) $HOCH_2-\overset{OH}{\underset{|}{C}}(CH_3)_2$

to $-CO_2H$ to $-CHO$ to $-CHO$ to HCHO   to ketone

7.69   (a) $CH_3CH_2\overset{O}{\underset{O}{\overset{||}{\underset{||}{S}}}}CH_2CH_3$ (b) (diagram) $S\!=\!O$ (c) $CH_3CH_2O\overset{O}{\overset{||}{C}}CH_3$

(d) $CH_3CH_2\overset{O}{\overset{||}{S}}C_6H_5$ (e) (diagram) $-S^-\ Na^+$

7.70   (a) $C_6H_5Br \xrightarrow{Mg} C_6H_5MgBr \xrightarrow[(2)\ H_2O,\ H^+]{(1)\ CO_2} C_6H_5CO_2H$

(b) $C_6H_5CH_2Br \xrightarrow{OH^-} C_6H_5CH_2OH \xrightarrow{CrO_3 \cdot pyridine} C_6H_5CHO$

$\downarrow Mg$

$C_6H_5CH_2MgBr$

$$\overset{OH}{\underset{|}{C_6H_5CHCH_2C_6H_5}} \xrightarrow{heat} \underline{trans}\text{-}C_6H_5CH{=}CHC_6H_5$$

(c) $(\underline{R})\text{-}\overset{OH}{\underset{|}{CH_3CHCH_2CH_3}} \xrightarrow[ether]{SOCl_2} (\underline{R})\text{-}\overset{Cl}{\underset{|}{CH_3CHCH_2CH_3}} \xrightarrow[S_N2]{OH^-} (\underline{S})\text{-}\overset{OH}{\underset{|}{CH_3CHCH_2CH_3}}$

7.71 (a) $CH_3CH_2\overset{CH_3}{\underset{|}{CHOH}} \xrightarrow{HBr} CH_3CH_2\overset{CH_3}{\underset{|}{CHBr}} \xrightarrow{Mg} CH_3CH_2\overset{CH_3}{\underset{|}{CHMgBr}}$

$$\xrightarrow[\text{(2) } H_2O,\ H^+]{\text{(1) HCHO}} CH_3CH_2\overset{CH_3}{\underset{|}{CHCH_2OH}} \xrightarrow{Na} CH_3CH_2\overset{CH_3}{\underset{|}{CHCH_2O^-}} \xrightarrow{CH_3I}$$

(b) $(CH_3)_3COH \xrightarrow{K} (CH_3)_3CO^-\ K^+ \xrightarrow{CH_3CH_2Br}$

(c) $(CH_3)_2CHOH \xrightarrow{HBr} (CH_3)_2CHBr \xrightarrow{Mg} (CH_3)_2CHMgBr$

$$\xrightarrow[\text{(2) } H_2O,\ H^+]{\text{(1) } CH_3CHO} (CH_3)_2CH\overset{OH}{\underset{|}{CHCH_3}} \xrightarrow{CrO_3 \atop H_2SO_4}$$

(d) $\overset{CH_2}{\overset{\diagup\diagdown}{CH_2CHCH_2OH}} \xrightarrow{PBr_3} \overset{CH_2}{\overset{\diagup\diagdown}{CH_2CHCH_2Br}} \xrightarrow{Mg} \overset{CH_2}{\overset{\diagup\diagdown}{CH_2CHCH_2MgBr}}$

$$\xrightarrow[\text{(2) } H_2O,\ H^+]{\text{(1) } CH_3CH_2CHO} \overset{CH_2}{\overset{\diagup\diagdown}{CH_2CHCH_2}}\overset{OH}{\underset{|}{CHCH_2CH_3}} \xrightarrow{SOCl_2}$$

(e) $CH_3CH_2OH \xrightarrow{HBr} CH_3CH_2Br \xrightarrow{Mg} CH_3CH_2MgBr$

$$\xrightarrow[\text{(2) } H_2O,\ H^+]{\text{(1) } CH_3\overset{O}{\overset{\|}{C}}CH_3} (CH_3)_2\overset{OH}{\underset{|}{C}}CH_2CH_3 \xrightarrow[heat]{H_2SO_4}$$

(f) $(CH_3)_2CHCH_2OH \xrightarrow{PBr_3} (CH_3)_2CHCH_2Br \xrightarrow{Mg}$

$(CH_3)_2CHCH_2MgBr \xrightarrow[\text{(2) } H_2O,\ H^+]{\text{(1) } \overset{\displaystyle O}{\overset{\diagup\diagdown}{CH_2CH_2}}}$

(g) $[(CH_3)_2CHCH_2MgBr \text{ from (f)}] \xrightarrow[\text{(2) } H_2O,\ H^+]{\text{(1) } CH_3CHO}$

$(CH_3)_2CHCH_2\overset{\displaystyle OH}{\underset{\displaystyle |}{C}}HCH_3 \xrightarrow{HBr} (CH_3)_2CHCH_2CHBrCH_3 \xrightarrow{CN^-}$

(h) $(CH_3)_2CHOH \xrightarrow{Na} (CH_3)_2CHO^-\ Na^+ \xrightarrow{\overset{\displaystyle O}{\overset{\diagup\diagdown}{CH_2CH_2}}} (CH_3)_2CHOCH_2CH_2OH \xrightarrow{SOCl_2}$

(i) $CH_3CH_2CH_2OH \xrightarrow{HBr} CH_3CH_2CH_2Br \xrightarrow{Mg} CH_3CH_2CH_2MgBr$

$\xrightarrow[\text{(2) } H_2O,\ H^+]{\text{(1) } (CH_3)_2CHCHO} CH_3CH_2CH_2\overset{\displaystyle OH}{\underset{\displaystyle |}{C}}HCH(CH_3)_2 \xrightarrow{HNO_3}$

7.72   (a) $CH_2{=}CHCH_2OH \xrightarrow{PCl_3} CH_2{=}CHCH_2Cl \xrightarrow{\text{excess NaSH}}$

$CH_2{=}CHCH_2SH \xrightarrow{I_2} \text{product}$

7.73 $\underset{\displaystyle CH_2{=}CH}{\overset{\displaystyle H{-}\overset{\displaystyle ..}{O}:}{}} \longrightarrow CH_2{-}CH \overset{\displaystyle H \quad \overset{..}{\underset{..}{O}}}{} $ (This reaction is probably not intramolecular as

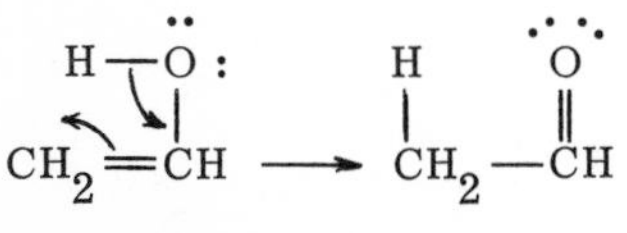

shown here. The proton may be lost to another molecule of $CH_2{=}CHOH$, which

may lose its OH proton to yet another molecule.)

7.74   (a) The cis-compound cannot undergo intramolecular S_N2 displacement of the

chlorine to yield an epoxide.

(b)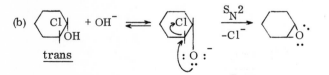

trans

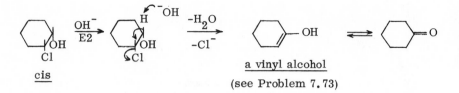

a vinyl alcohol

(see Problem 7.73)

7.75   (b) is the stronger acid because the Cl is electron-withdrawing and can help dis-
perse the negative charge in the anion by the inductive effect.

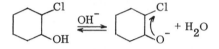

7.76   (d). Only (a) and (d) undergo a rapid acid-base reaction with a Grignard reagent.

$$ROH + CH_3MgI \longrightarrow ROMgI + CH_4$$

Of these two alcohols, (a) undergoes rapid reaction with Lucas reagent because it
is a 3° alcohol. (d) undergoes slow reaction with Lucas reagent because it is a
less-reactive 1° alcohol.

7.77   (a)   (b)   (c)

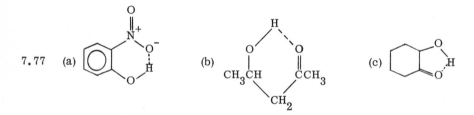

7.78   It cannot undergo intramolecular hydrogen bonding because the H and O are too far
apart.

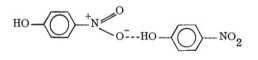

7.79   (a) $C_6H_5CH_3$ $\xrightarrow{NBS}$ $C_6H_5CH_2Br$ $\xrightarrow{Mg}$ $C_6H_5CH_2MgBr$ $\xrightarrow[\text{(2) } H_2O, H^+]{\text{(1) } CH_3CHO}$

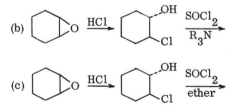

(d) CH_2CH_2 (epoxide) $\xrightarrow[\text{(2) } H_2O,\ H^+]{\text{(1) } C_6H_5O^-}$ $C_6H_5OCH_2CH_2OH \xrightarrow{SOCl_2}$

$C_6H_5OCH_2CH_2Cl \xrightarrow{C_6H_5O^-}$

(e) $(CH_3CH_2)_2C\overset{O}{\diagup\!\diagdown}C(CH_2CH_3)_2 \xrightarrow{H_2O,\ H^+}$

$(CH_3CH_2)_2\overset{\overset{\displaystyle OH}{|}}{C}-\overset{\overset{\displaystyle OH}{|}}{C}(CH_2CH_3)_2 \xrightarrow[\text{(pinacol rearrangement)}]{H^+}$

7.80   In a road-map problem, find your clue (in this case, the reaction of the Grignard reagent of B with CH_3CHO to yield 5-methyl-2-heptanol) and then work the problem backwards.

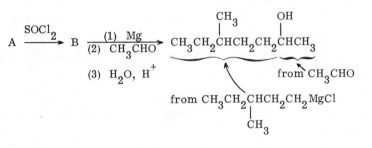

$A \xrightarrow{SOCl_2} B \xrightarrow[\substack{\text{(2) } CH_3CHO \\ \text{(3) } H_2O,\ H^+}]{\text{(1) Mg}} CH_3CH_2\overset{\overset{\displaystyle CH_3}{|}}{C}HCH_2CH_2\overset{\overset{\displaystyle OH}{|}}{C}HCH_3$

from CH_3CHO

from $CH_3CH_2\overset{\overset{\displaystyle}{}}{C}HCH_2CH_2MgCl$
$\qquad\qquad\;\; |$
$\qquad\qquad\; CH_3$

B is $CH_3CH_2\overset{\overset{\displaystyle CH_3}{|}}{C}HCH_2CH_2Cl$ and therefore A is $CH_3CH_2\overset{\overset{\displaystyle CH_3}{|}}{C}HCH_2CH_2OH$.

Chapter 8

Spectroscopy I:
Infrared and Nuclear Magnetic Resonance

Some Important Features

Electromagnetic radiation can be absorbed by organic compounds, and this absorption of energy results in increased energy of the molecules. Different compounds absorb electromagnetic radiation of different energy, and thus of different wavelength (λ) or frequency (ν). Radiation of shorter wavelength is of higher frequency and higher energy than radiation of longer wavelength.

Absorption of infrared radiation results in increased bond vibrations. Polar groups usually exhibit stronger peaks in the infrared spectrum than nonpolar groups. The usual positions of absorption are shown in Figure 8.7 in the text. The absorption positions (and peak appearances) of —OH, —NH and —NH$_2$ (a double peak), —CO$_2$H, and C$=$O are particularly distinctive.

An nmr spectrum results from the change in spin states of hydrogen nuclei (protons). The position of absorption by a proton (the <u>chemical shift</u>) is affected by neighboring atoms and by other groups in the molecule. A nearby electronegative atom results in deshielding, and absorption is observed farther downfield.

$$\longleftarrow \quad \begin{array}{c} \text{downfield} \\ \text{(deshielded)} \end{array} \qquad \begin{array}{c} \text{upfield} \\ \text{(shielded)} \end{array} \quad \longrightarrow$$

Other groups in the molecule affect the position of absorption by <u>anisotropic effects</u>. Absorption by protons attached to aromatic rings, to aldehyde carbonyl groups, and to $\underline{sp}^2$ carbons of alkenes is observed downfield. Figure 8.29 in the text shows some typical positions of proton absorption in nmr spectra.

Recognizing the equivalence or nonequivalence of protons in nmr spectroscopy is important in structure determination. We suggest that you review Section 8.9A.

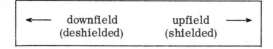

The area under the signal for a proton is proportional to the number of protons giving rise to that signal. In the above example, we observe two principal signals: the CH_3 signal and the CH_2 signal. The area ratio is $6:4$, or $3:2$.

Absorption peaks in the nmr spectrum may be split into multiple peaks. A group of equivalent protons absorb radio waves at the same position of the nmr spectrum and do not split the signals of each other. Vicinal protons that are not equivalent to the proton in question do split the signal of the proton.

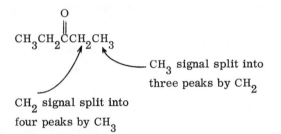

CH$_3$ signal split into
three peaks by CH$_2$

CH$_2$ signal split into
four peaks by CH$_3$

If a signal is split by a group of vicinal protons equivalent to each other, then the $\underline{n+1}$ rule is followed. In the above example, CH_3 is split by CH_2 (two protons equivalent to each other but not equivalent to the CH_3 protons). The splitting pattern for CH_3 is $2+1$, or a triplet.

Another topic covered in the nmr discussion is <u>chemical exchange</u>, which results in no splitting of (or by) an OH or NH proton.

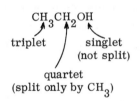

triplet   singlet
               (not split)

quartet
(split only by CH$_3$)

Section 8.13 in the text is an introduction to determination of structure from the molecular formula and spectral data. This section should be studied carefully, as well as the discussions in the answers to Problem 8.45 (after you have solved them or tried to solve them).

Before studying spectra in detail for clues about a structure, first examine the molecular formula and the spectra for obvious clues.

1. Does the structure contain an aromatic ring? Check the nmr spectrum for aryl protons.

2. Does the structure contain $C{=}O$, NH or OH, NH_2, or CO_2H? Check the infrared spectrum for these distinctive peaks.

3. How many rings or double bonds does the structure contain? Check the molecular formula. Remember that the $C{=}O$ group is a site of unsaturation. Butanone has the

general formula C_4H_8O, or $C_nH_{2n}O$. A phenyl group contains three sites of unsatura-tion <u>and</u> a ring. $C_6H_5CH_2CH_3$ has the general formula $C_nH_{2\underline{n}-6}$.

4. Does the structure contain an ethyl group bonded to an electronegative atom? Check the nmr spectrum for an upfield triplet and a downfield quartet.

Answers to Problems

8.17 (a) $3.33\,\mu m$ (b) $5.68\,\mu m$ (c) $1785\ cm^{-1}$ (d) $1220\ cm^{-1}$ (e) 3×10^{-5} MHz

(a) through (d) are solved by the following equations:

$$\text{wavenumber in } cm^{-1} = \frac{1}{\lambda \text{ in cm}}$$
$$1\,\mu m = 10^{-4}\ cm$$

8.18 (a) 8 nm   (b) $3000\ cm^{-1}$ (c) infrared   (d) 60 Hz

In each case, the electromagnetic radiation of shorter wavelength has the higher energy.

8.19 The higher-energy absorption (higher frequency, shorter wavelength) is on the left side.

8.20 (a) $C{=}O$ (b) $C{=}C{-}Cl$ (c) $O{-}H$. In each case, the vibration that results in the greater change in dipole moment gives stronger absorption.

8.21 (a) $CH_3CH_2CH_2NH_2$ shows NH absorption, while $CH_3CH_2CH_2N(CH_3)_2$ does not
(b) $CH_3CH_2CH_2CO_2H$ shows OH absorption, while $CH_3CH_2CH_2CO_2CH_3$ does not
(c) $CH_3CH_2CO_2CH_3$ shows C$-$O absorption, while $CH_3CH_2\underset{\underset{O}{\|}}{C}CH_3$ does not

8.22

As the reaction proceeds, the OH absorption of the organic extract gradually dis-appears. The carbonyl absorption of cyclohexanone is not useful in this case be-cause it appears early in the course of the reaction.

8.23 Any cyclic ether containing five carbon atoms is consistent with the data.

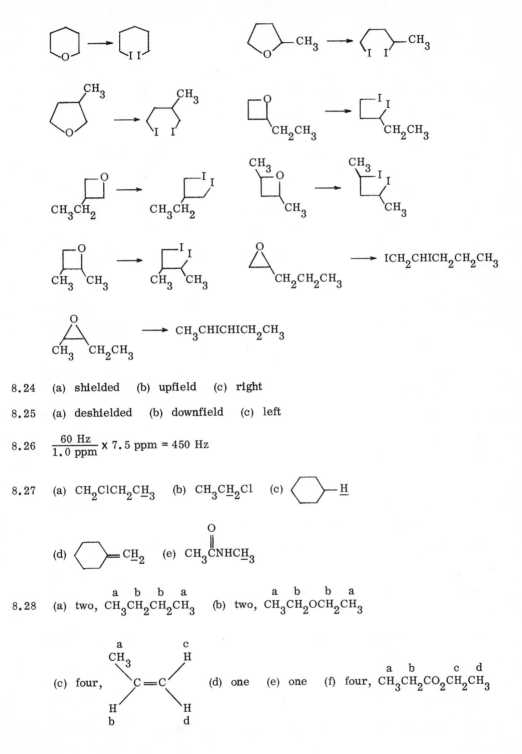

8.24 (a) shielded   (b) upfield   (c) right

8.25 (a) deshielded   (b) downfield   (c) left

8.26 $\dfrac{60\ Hz}{1.0\ ppm}$ × 7.5 ppm = 450 Hz

8.27 (a) $CH_2ClCH_2C\underline{H}_3$ (b) $CH_3C\underline{H}_2Cl$ (c) ⬡—$\underline{H}$

(d) ⬡=$C\underline{H}_2$ (e) $CH_3\overset{\overset{O}{\|}}{C}NHC\underline{H}_3$

8.28 (a) two, $\overset{a}{C}H_3\overset{b}{C}H_2\overset{b}{C}H_2\overset{a}{C}H_3$ (b) two, $\overset{a}{C}H_3\overset{b}{C}H_2O\overset{b}{C}H_2\overset{a}{C}H_3$

(c) four, $\underset{\underset{b}{H}}{\overset{\overset{a}{CH_3}}{C}}=\underset{\underset{d}{H}}{\overset{\overset{c}{H}}{C}}$ (d) one   (e) one   (f) four, $\overset{a}{C}H_3\overset{b}{C}H_2CO_2\overset{c}{C}H_2\overset{d}{C}H_3$

(g) four, $\overset{a}{C}H_3\overset{b}{C}HCl\overset{c}{C}H_2\overset{d}{C}H_3$ (h) same as (g)   (i) three, $(\overset{a}{C}H_3)_2\overset{b}{C}H\overset{b}{N}\overset{a}{C}H(CH_3)_2$
$\overset{|}{\underset{c}{C}H_3}$

(j) four, $Br\overset{a}{C}H_2\overset{b}{C}H_2\overset{c}{C}H(\overset{d}{C}H_3)_2$ (k) four,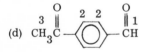

(l) two, $\overset{a}{C}H_3O$—⟨benzene⟩—$\overset{}{O}CH_3$ with b b above and b b below

(m) four, $\overset{a}{C}H_3O$—⟨benzene⟩—$\overset{d}{C}H_3$ with b c above and b c below

8.29   (a) two   (b) two   (c) four   (d) one   (e) one   (f) four   (g) four

(h) four   (i) three   (j) four   (k) four   (l) two   (m) four

In each case, the number of principal signals would probably be the same as the number of nonequivalent protons.

8.30   Tetraethylsilane would exhibit a triplet and a quartet, not a singlet, in the nmr spectrum. The triplet would fall in the same general region of a spectrum as the absorption of other organic compounds.

8.31   (a) $\overset{6}{(CH_3)_2}\overset{2}{N}CH_2\overset{3}{C}H_3$ (b) $\overset{6}{(CH_3}\overset{4}{C}H_2)_2\overset{1}{C}HCl$ (c) $\overset{3}{(CH_3)_3}\overset{}{C}O\overset{1}{C}H_3$

(d) $\overset{3}{C}H_3\overset{O}{\overset{||}{C}}$—⟨benzene⟩—$\overset{O}{\overset{||}{C}}\overset{1}{H}$ with 2 2 above benzene

8.32   (d) and (e). None of the other possibilities have nonequivalent protons in the ratio of 3:1. Instead, the following ratios are observed:

(a) 3:1:1:1   (b) 3:1:1:1   (c) 3:2:2   and   (f) 1:1

8.33   Numbers represent multiplicity:

(a) $\overset{}{CCl_3}\overset{3}{C}H_2\overset{3}{C}H_2Br$ (b) $\overset{2}{(CH_3)_2}\overset{7}{C}H\overset{1}{C}O_2CH_3$ (c) $\overset{1}{(CH_3)_3}\overset{1}{C}OCH_3$

(d) 3⟨benzene⟩—$\overset{7}{C}H(\overset{2}{C}H_3)_2$ with 3 2 below benzene

(The signals for the aryl protons may not actually show all the indicated splitting.)

8.34   (a) $\overset{3}{C}H_3\overset{4}{C}H_2\overset{}{C}O_2\overset{1}{C}H_3$ (areas, 3:2:3)   (b) $\overset{1}{C}H_3OCH_2CH_2OCH_3$ (areas, 3:2)

(c) $\overset{1}{C}H_3\overset{O}{\overset{\|}{C}}\overset{4}{C}H_2\overset{3}{C}H_3$ (areas, 3:2:3)   (d) $\overset{1}{C}H_3O\!\!-\!\!\overset{2\ 2}{\bigcirc}\!\!-\!\!Cl$ (areas, 3:2:2)

(e) $Cl_2\overset{3}{C}HCH_2\overset{2}{B}r$ (areas, 1:2)

8.35 Cl_2CHCH_3
(a)   (b)

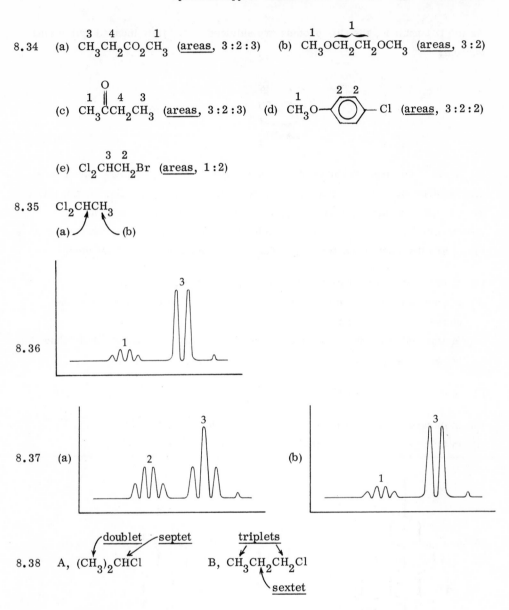

8.36

8.37   (a)   (b)

8.38   A, $(CH_3)_2CHCl$ ⟨doublet, septet⟩   B, $CH_3CH_2CH_2Cl$ ⟨triplets, sextet⟩

8.39   (a) $CH_3CH_2CH_2CH\!=\!CH_2$ shows a large number of peaks, while
         $(CH_3)_2C\!=\!C(CH_3)_2$ shows a singlet
       (b) CH_3CH_2CHO shows three principal peaks (including an offset peak for an
         aldehyde proton), while CH_3COCH_3 shows a singlet
       (c) CH_3COCH_3 shows one singlet, while $CH_3CO_2CH_3$ shows two singlets
       (d) $CH_3CH_2\!-\!C_6H_5$ shows a quartet and a triplet upfield, while
         $\underline{p}\text{-}CH_3\!-\!C_6H_4\!-\!CH_3$ shows a singlet upfield

8.40   The outside protons, like those of benzene, are deshielded and absorb downfield in
       the aromatic region. The inner protons are highly <u>shielded</u> and absorb upfield.
       (This absorption is observed upfield even of TMS.)

8.41   As in [18]annulene, the CH_3 protons are shielded by the field induced by the ring current.

8.42   (a) seven   (b) aryl, doublet (area 2) plus triplet (area 1), but may appear as a singlet (area 3); CH_3, singlet (area 6); NH, singlet (area 1); CH_2, singlet (area 2); CH_3CH_2, triplet (area 4) plus quartet (area 3)   (c) NH absorption (~ 3500 cm^{-1}), C—H absorption (~ 3000 cm^{-1}), and C=O absorption (~ 1700 cm^{-1}) are the most characteristic peaks we have discussed in this chapter.

8.43   (a) In infrared, 1-propanol shows OH; propylene oxide does not.

   (b) In nmr, diisopropyl ether shows a doublet and a septet; di-n-propyl ether shows two triplets and a sextet.

   (c) In nmr, ethanol shows one singlet, one triplet, and one quartet; 1,2-ethanediol shows two singlets.

   (d) In infrared, ⟨ ⟩NH shows NH; ⟨ ⟩NCH$_3$ does not.

   (e) In infrared, ethanol shows OH; ethyl chloride does not.

   (f) In infrared, acetic acid shows broad OH; acetone does not show OH.

8.44   (a)   (b)   (c)

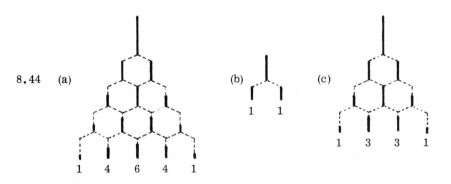

8.45   (a) The infrared spectrum shows NH_2 near 3400 cm^{-1}. The formula may thus be rewritten $C_4H_9NH_2$, and we can see that no ring or double bond is present. The nmr spectrum shows two types of proton (neither with vicinal protons). The NH_2 accounts for one peak (area 2). The other must be a t-butyl group (area 9). The compound is $(CH_3)_3CNH_2$.

   (b) The infrared spectrum shows carbonyl absorption, while the nmr shows aryl protons (area 5). This information alone tells us two pieces of the structure: C_6H_5— and C=O. The other two peaks in the nmr spectrum are not split, and therefore are not from vicinal protons. From their areas and from the remainder of the molecular formula (C_2H_5), we deduce the presence of a CH_2 group and a CH_3 group. The compound is $C_6H_5CH_2CCH_3$.

(c) The infrared spectrum shows an OH group, while the nmr spectrum shows aryl protons and the OH proton (the singlet at 3.8 ppm). The triplet at 4.3 ppm must arise from a proton adjacent to the OH (from its chemical shift) and must be attached to a CH_2 group (because of its multiplicity). The CH_2 group (quintet) is also attached to a methyl group (triplet at 0.5 ppm).

$$C_6H_5- \qquad CH_3CH_2CHOH$$

The compound is $C_6H_5CHCH_2CH_3$.
OH

(d) This one is easy: $C_6H_5CH_3$.

(e) This compound does not have a phenyl group (see the nmr spectrum). The formula $(C_nH_{2n-2}O_4)$ shows two sites of unsaturation or rings. At least one of these is a $\overline{C=O}$ group (see the infrared spectrum).

Look at the nmr spectrum in detail. The sum of the area ratios under the peaks is six, half the actual number of protons in the molecular formula. Therefore, we double the area numbers (4:2:6) to tell us the actual number of protons giving rise to each signal. Then the nmr spectrum shows two CH_3CH_2 groups and a CH_2 group. Because of the chemical shifts, we can assume the two ethyl groups are attached to oxygen atoms. So far, we have accounted for the following fragments:

O
‖
—C— CH_3CH_2O- CH_3CH_2O- $-CH_2-$ plus C and O

There is no way these fragments can be pieced together to form a ketone or a diether. Instead, the structure contains two <u>ester</u> groups. (See the infrared spectrum for corroboration.)

O   O
‖   ‖
$-COCH_2CH_3$ $-COCH_2CH_3$ $-CH_2-$

We can now put these pieces together to arrive at the structure of the compound: $CH_2(COCH_2CH_3)_2$.
‖
O

(f) The formula $(C_nH_{2n}O)$ indicates a ring or a double bond. The infrared spectrum shows no $\overline{C=O}$ or OH groups; therefore, the compound must be an ether.

The nmr spectrum gives us the following information (after we double the areas to fit the formula).

(1) A quintet from two equivalent H's: $—CH_2C\underline{H}_2CH_2—$

(2) A triplet from four equivalent H's: $—C\underline{H}_2CH_2C\underline{H}_2—$

(3) No alkene protons

The only formula that fits these facts and the chemical shifts is a cyclic ether:

$$\begin{array}{ccc} CH_2 & — & CH_2 \\ | & & | \\ O & —— & CH_2 \end{array}$$

(g) This compound exhibits a complex nmr spectrum that cannot be entirely interpreted. For example, the downfield "quartet" is actually a triplet superimposed on a quartet. (Your clue comes from the relative heights of the peaks, which are abnormal for a quartet.)

By using the formula and the infrared spectrum, as well as the nmr spectrum, we can deduce that the downfield quartet comes from an ethyl ester group.

$$—CO_2C\underline{H}_2CH_3 \qquad \text{quartet}$$

The quintet at 2.0 ppm arises from a proton that is split by four vicinal protons.

$$\text{quintet} \qquad C\underline{H}_3C\underline{H}_2\overset{\displaystyle \overset{Br}{|}}{C\underline{H}}— \qquad \text{triplet that overlaps quartet}$$

The upfield group (actually two overlapping triplets) is from two methyl groups, both shown in the preceding fragments. Putting all the data together, we arrive at the following formula for the unknown compound:

$$CH_3CH_2CHBrCO_2CH_2CH_3.$$

Chapter 9

Alkenes and Alkynes

Some Important Features

The principal reactions discussed in Chapter 9 are additions of reagents to carbon-carbon double bonds. These addition reactions can be grouped together according to the types of intermediate formed. Addition of HX or H_2O usually proceeds by way of the more stable carbocation to yield Markovnikov products.

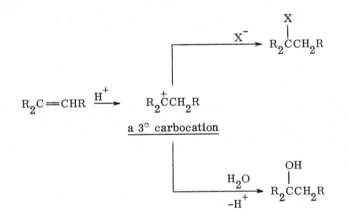

A carbocation undergoes rearrangement if a more stable carbocation will result. In predicting products from carbocation reactions, always inspect the structure of the intermediate carbocation to see if it can rearrange to a more stable intermediate.

1,4-Addition reactions yield allylic cations and often result in mixtures of products. To predict the products, look at the more stable carbocations and at their resonance structures. The principal products arise from the major contributors.

Addition reactions of X_2 or $Hg(O_2CCH_3)_2$ proceed through bridged intermediates, which result in <u>anti</u>-addition. No rearrangements occur, but Markovnikov's rule is still followed.

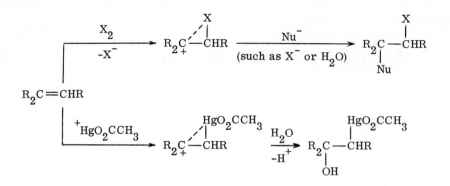

Other types of addition reactions are those of HBr with peroxides (a free-radical, anti-Markovnikov reaction); BH_3 (which yields what appear to be anti-Markovnikov products); singlet methylene (<u>syn</u>-addition); and H_2 (<u>syn</u>-addition).

Oxidation reactions of alkenes can lead to a variety of products, depending on the reagent and on the degree of alkene substitution.

$$R_2C=CHR \xrightarrow[\text{syn-addition}]{\substack{\text{cold KMnO}_4 \text{ solution} \\ \text{or (1) OsO}_4, \quad \text{(2) Na}_2\text{SO}_3}} \underset{\overset{|}{\text{OH}} \quad \overset{|}{\text{OH}}}{R_2C-CHR}$$

$$R_2C=CHR \xrightarrow{\text{hot KMnO}_4 \text{ solution}} R_2C=O + HOCR \overset{O}{\overset{\|}{}}$$

Electrocyclic reactions (Section 9.16) and Diels-Alder reactions (Section 9.17) are different types of addition reactions of conjugated polyenes. In these reactions, as in 1,4-addition reactions, the position of the double bond in the product is different from the starting positions.

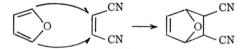

<u>The two end carbons of the diene</u>
<u>attack the carbon-carbon</u>
<u>double bond of the dienophile.</u>

Reminders

In an addition reaction that proceeds by way of a carbocation (or a bridged intermediate with carbocation character), check the relative stabilities of all possible intermediates. The nucleophile will attack predominantly the most positive carbon.

$$\frac{2° \quad 3° \quad \text{allylic and benzylic}}{\text{increasing carbocation stability}} \longrightarrow$$

Weak nucleophiles (like water, an alcohol, or an anion) can attack a carbocation.

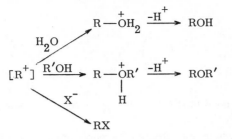

Reactions that proceed through true carbocations are not stereospecific. However, anti-addition reactions, which proceed by way of bridged ions, and syn-addition reactions are stereospecific.

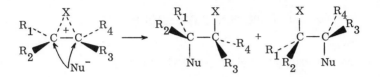

Answers to Problems

9.42   (a) 2-butene   (b) 2-methyl-2-butene   (c) 1-methyl-1-cyclohexene

(d) 3-methyl-1-butyne   (e) 2-methyl-3-penten-1-ol   (f) 1,3-butadiene

(g) 3-propyl-2-hexene   (h) 1,3-cyclohexadiene

9.43   (a) 3-bromo-1-propene   (b) 3-methyl-2-buten-1-ol   (c) propenoic acid

(d) 5,5-dichloro-1,3-cyclohexadiene

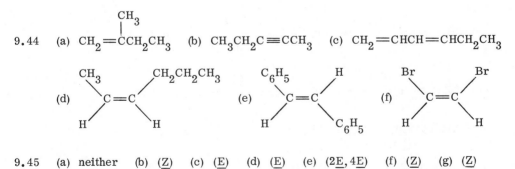

9.44   (a) $CH_2{=}\overset{\underset{\displaystyle |}{CH_3}}{C}CH_2CH_3$ (b) $CH_3CH_2C{\equiv}CCH_3$ (c) $CH_2{=}CHCH{=}CHCH_2CH_3$

9.45   (a) neither   (b) (Z)   (c) (E)   (d) (E)   (e) (2E, 4E)   (f) (Z)   (g) (Z)

(h) (E)

Compound (a) has no geometric isomers; therefore, the (E) and (Z) system is not applicable.

In (e), consider each double bond separately. First, compare the groups attached to carbons 2 and 3. Second, compare the groups attached to carbons 4 and 5. We have circled the higher-priority groups.

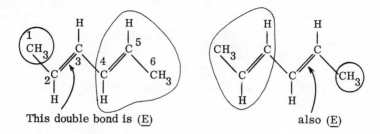

This double bond is (E)   also (E)

In (f), first decide which four atoms or groups are attached to the double-bond carbons. The higher-priority groups make up the rest of the ring, which must be on the same side of the double bond, or (Z).

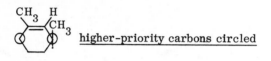

higher-priority carbons circled

Compound (g) represents a similar case and is also (Z).

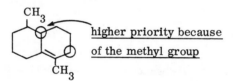

higher priority because
of the methyl group

9.46   (a) (1), because (2) would yield predominantly 2-butene (Saytseff rule).

(b) (1), because (2) would also yield a large proportion of 3-methyl-1-cyclohexene.

9.47   (a) $CH_3C \equiv CH \xrightarrow[\text{deactivated Pd}]{H_2} CH_2CH = CH_2$

(b) $CH_3C \equiv CH \xrightarrow{CH_3MgI} CH_3C \equiv CMgI \xrightarrow[\text{(2) } H_2O, H^+]{\text{(1) } \bigcirc = O} \text{ product}$

(c) $CH_3C \equiv CH \xrightarrow{NaNH_2} CH_3C \equiv C^- Na^+ \xrightarrow{CH_3I} CH_3C \equiv CCH_3$

9.48   (a) $CHCl = CClCH_2CH_2CH_3$ (b) $CHCl_2CCl_2CH_2CH_2CH_3$

(c) $CH \equiv CCH_2CH_2CH_3 \xrightarrow{HCl} CH_2 = CClCH_2CH_2CH_3$

Markovnikov product

$\xrightarrow{HCl} CH_3CCl_2CH_2CH_2CH_3$

Markovnikov product

(d) $BrMgC \equiv CCH_2CH_2CH_3 + C_6H_6$

(e) $CH\equiv CCH_2CH_2CH_3 \xrightarrow[-NH_3]{NH_2^-} {}^-C\equiv CCH_2CH_2CH_3$

$\xrightarrow[-I^-]{CH_3I} CH_3C\equiv CCH_2CH_2CH_3$

9.49 (a) $CH_3CHICH_2CH_2CH_3$

(b) $CH_2=CHCH=CHCH_3 \xrightarrow[Markovnikov]{H^+}$

$$\left[CH_3\overset{+}{C}HCH=CHCH_3 \longleftrightarrow CH_3CH=CH\overset{+}{C}HCH_3 \right]$$
$$\underline{equivalent}$$

$$+ \left[CH_2=CHCH_2\overset{+}{C}HCH_3 \right]$$
$$\underline{not\ resonance\text{-}stabilized}$$

$$+ \left[CH_2=CH\overset{+}{C}HCH_2CH_3 \longleftrightarrow \overset{+}{C}H_2CH=CHCH_2CH_3 \right] \xrightarrow{I^-}$$

$CH_3CH=CHCHICH_3 + CH_2=CHCHICH_2CH_3 + $ some $ICH_2CH=CHCH_2CH_3$

(c) As in (b), first determine the more stable carbocation intermediates and their resonance structures; then, determine the products:

$(CH_3)_2CICH=CH_2 + CH_2=CCHICH_3 + $ some $(CH_3)_2C=CHCH_2I$
$\qquad\qquad\qquad\qquad\quad |$
$\qquad\qquad\qquad\qquad CH_3$

and $\ ICH_2C=CHCH_3$
$\qquad\qquad\quad |$
$\qquad\qquad\ CH_3$

(d) $(CH_3)_2CICH(CH_2)_3CH_3 + $ some $(CH_3)_3CCH_2CHICH_2CH_2CH_3$
$\qquad\quad |$
$\qquad\ CH_3$

9.50 (a) $CH_2=CHCl$, because it leads to the more stabilized carbocation (relatively speaking) $CH_3\overset{+}{C}H \rightarrow Cl$ and not $Cl \leftarrow \overset{+}{C}H \rightarrow CH_2Cl$.

(b) $CH_3CH=C(CH_3)_2$, because it yields a more stabilized tertiary carbocation.

9.51 (a), (c), (b). The order is based upon the stabilities of the carbocations formed by protonation.

9.52 (a) $\left[(CH_3)_2CH^+ \right] \longrightarrow (CH_3)_2CHOSO_3H$

(b) $\left[(CH_3)_3C^+ \right] \longrightarrow (CH_3)_3COSO_3H$

(c) $\left[CH_3\overset{+}{C}HCH_2CH_3 \right]$ $\longrightarrow$ $CH_3\overset{\underset{\displaystyle |}{OSO_3H}}{C}HCH_2CH_3$

9.53   (a) $(CH_3CH_2)_2\overset{\underset{\displaystyle |}{OH}}{C}CH_3$ (b) $(CH_3)_3COCH_2CH_3$ (c) $(CH_3)_2\overset{\underset{\displaystyle CH_3}{\overset{\displaystyle OH}{|}}}{C}CHCH_2CH_2CH_3$

(d) $CH_3\overset{\underset{\displaystyle |}{OH}}{C}HCH_2CH_3$ (e)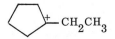

In (e), the alkene is protonated and then attacked by the nucleophilic $CF_3CO_2^-$.

$$\text{(cyclohexene)}=CH_2 \xrightarrow{H^+} \text{(cyclohexyl)}\overset{+}{{}}-CH_3 \xrightarrow{CF_3CO_2^-} \text{product}$$

9.54 $(CH_3)_3CCH=CHCH_2CH_3 \xrightarrow{H^+} \left[(CH_3)_3C\overset{+}{C}HCH_2CH_2CH_3 \right] \longrightarrow$

$\left[(CH_3)_2\overset{\underset{\displaystyle |}{\overset{\displaystyle CH_3}{\overset{+}{}}}}{C}CHCH_2CH_2CH_3 \right] \xrightarrow{H_2O} (CH_3)_2\overset{\underset{\displaystyle \overset{+}{O}H_2}{\overset{\displaystyle CH_3}{|}}}{C}CHCH_2CH_2CH_3 \underset{\displaystyle \longleftarrow}{\overset{\displaystyle -H^+}{\longrightarrow}}$

$(CH_3)_2\overset{\underset{\displaystyle \overset{\displaystyle |}{OH}}{\overset{\displaystyle CH_3}{|}}}{C}CHCH_2CH_2CH_3$

9.55   (a) $(CH_3)_3C\overset{\underset{\displaystyle |}{OH}}{C}HCH_2CH_3 + (CH_3)_3CCH_2\overset{\underset{\displaystyle |}{OH}}{C}HCH_3$

(b) $(CH_3)_2CH\overset{\underset{\displaystyle |}{OCH_2CH_3}}{C}HCH_3$ (c) (cyclopentane)$\overset{OH}{{}}-CH_3$ (d) $(CH_3)_2CHCH_2CH_2OH$

9.56   (a) (cyclopentane)$\overset{\underset{\displaystyle CH_2CH_3}{\overset{\displaystyle Br}{}}}{{}}$ (b) (cyclopentane)$-CHBrCH_3$

Reaction (a) is an ionic reaction that proceeds through an intermediate tertiary carbocation:

(cyclopentane)$\overset{+}{{}}-CH_2CH_3$

Reaction (b) is a free-radical reaction that proceeds through an intermediate tertiary free radical:

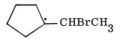

9.57   HCl (bond dissociation energy, 102 kcal/mole) is not as easily broken into free radicals as is HBr (bond dissociation energy, 87 kcal/mole).

9.58   (a) (2R, 3R)- and (2S, 3S)-$CH_3CHClCHClCH_2CH_3$

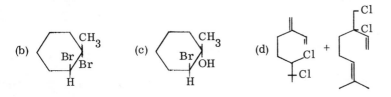

In (a), (b), and (c), the stereochemistry is a result of anti-addition (see Section 9.11). In (d), the predominant products are those from intermediates with tertiary carbocation character. For example:

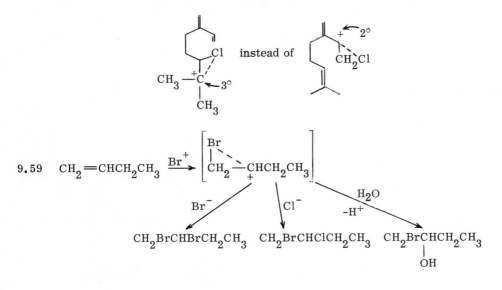

9.59 CH_2=$CHCH_2CH_3$ $\xrightarrow{\text{Br}^+}$ [$\overset{\text{Br}}{CH_2}$—$\overset{+}{CH}CH_2CH_3$]

$\swarrow$ Br$^-$ $\downarrow$ Cl$^-$ $\searrow$ $\underset{-H^+}{H_2O}$

$CH_2BrCHBrCH_2CH_3$ $CH_2BrCHClCH_2CH_3$ $CH_2BrCHCH_2CH_3$
                                                        $\underset{|}{OH}$

9.60   (a) H_2, Pt   (b) Br_2 + H_2O

9.61 CH_2=$\overset{\underset{\displaystyle CH_3}{|}}{C}$CH=$CH_2$ $\longrightarrow$ $\left[BrCH_2\overset{\underset{\displaystyle CH_3}{|}}{\overset{+}{C}}CH=CH_2 \longleftrightarrow BrCH_2\overset{\underset{\displaystyle CH_3}{|}}{C}=CH\overset{+}{C}H_2 \right]$

I   II

or $\left[CH_2=\overset{\underset{\displaystyle CH_3}{|}}{\underset{+}{C}}CHCH_2Br \longleftrightarrow \overset{+}{C}H_2\overset{\underset{\displaystyle CH_3}{|}}{C}=CHCH_2Br \right]$

III   IV

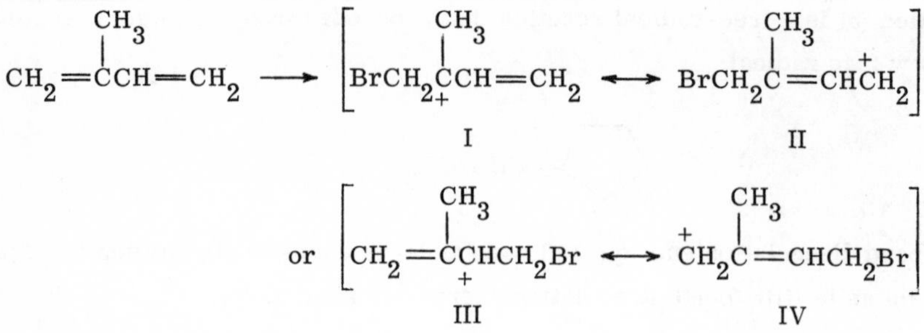

II and IV (primary carbocations) are not major contributors; therefore, the observed products are:

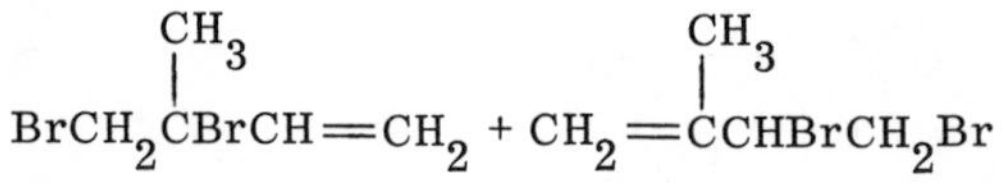

$BrCH_2\overset{\underset{\displaystyle CH_3}{|}}{C}BrCH=CH_2 + CH_2=\overset{\underset{\displaystyle CH_3}{|}}{C}CHBrCH_2Br$

(Note that II and IV would lead to the same product, and that *cis* and *trans* isomers are possible).

9.62 $(CH_3)_3\overset{+}{N}-\overset{-}{C}H_2$ $\xrightarrow{\text{heat}}$ $(CH_3)_3N: + :CH_2$

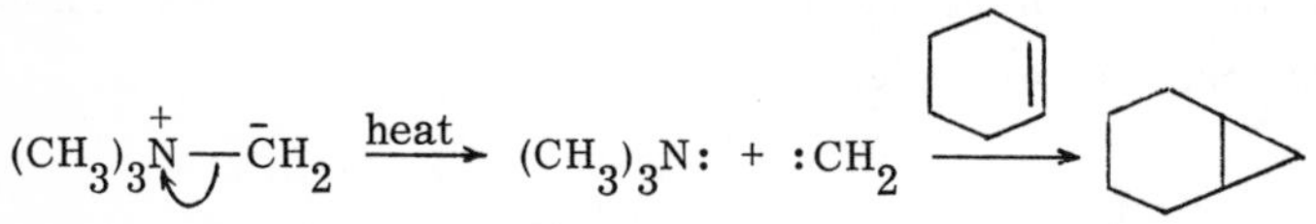

9.63   (a)

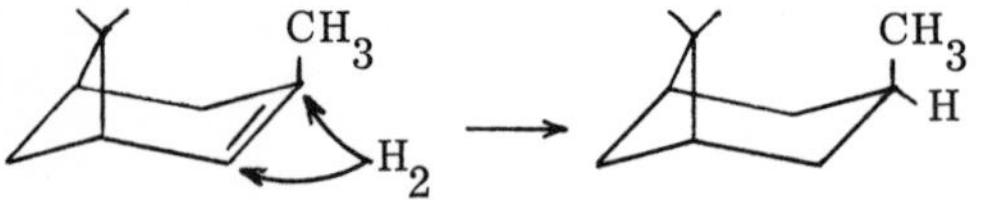

attack **trans** to bridge   **cis** product

(b) **cis** and **trans** because there is not enough hindrance for steric control of

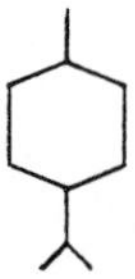

H_2 addition.

(c) 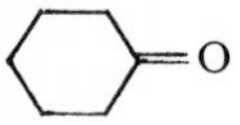 (d) 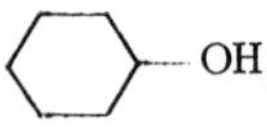 —OH   (e) **cis**-CH_3CH_2CH=$CHCH(CH_3)_2$

9.64   (a) 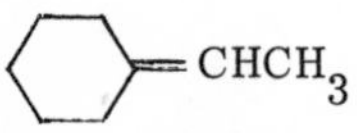 =$CHCH_3$ (most substituted)   (b) **trans**   (c) **(E)**

(d) CH_2=$CHCO_2CH_2CH_3$ (conjugated)

9.65   (a) 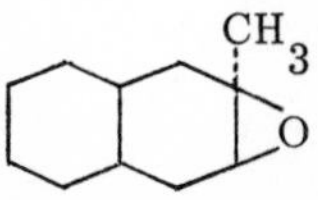   (b) 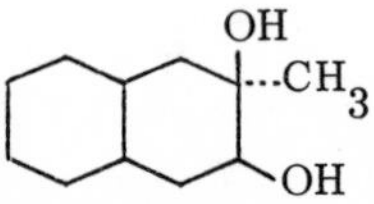 (**cis**-hydroxylation)

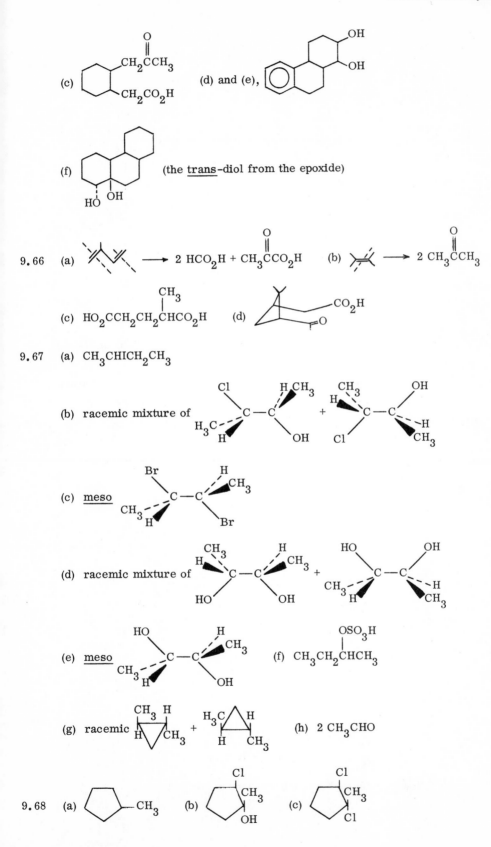

(c)

(d) and (e),

(the trans-diol from the epoxide)

(f)

9.66 (a) ⟶ 2 HCO_2H + CH_3CCO_2H (b) ⟶ 2 CH_3CCH_3

(c) $HO_2CCH_2CH_2CHCO_2H$ (d)

9.67 (a) $CH_3CHICH_2CH_3$

(b) racemic mixture of

(c) meso

(d) racemic mixture of

(e) meso   (f) $CH_3CH_2CHCH_3$

(g) racemic   (h) 2 CH_3CHO

9.68 (a)   (b)   (c)

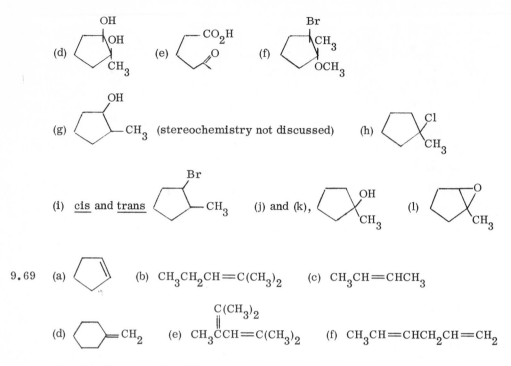

(d) [structure with OH, OH, CH₃]   (e) [structure with CO₂H, O]   (f) [structure with Br, CH₃, OCH₃]

(g) [structure with OH, CH₃] (stereochemistry not discussed)   (h) [structure with Cl, CH₃]

(i) <u>cis</u> and <u>trans</u> [structure with Br, CH₃]   (j) and (k), [structure with OH, CH₃]   (l) [structure with O, CH₃]

9.69   (a) [cyclopentene]   (b) $CH_3CH_2CH=C(CH_3)_2$ (c) $CH_3CH=CHCH_3$

(d) [cyclohexane]$=CH_2$ (e) $CH_3\overset{\overset{\displaystyle C(CH_3)_2}{\|}}{C}CH=C(CH_3)_2$ (f) $CH_3CH=CHCH_2CH=CH_2$

To solve this type of problem, redraw the products so that the carbonyl groups are close together (because two carbonyl groups result from one double bond). For example,

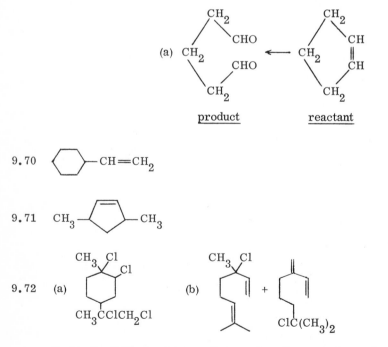

(a) [cyclic structure with CH₂, CHO, CH₂, CHO, CH₂]   ⟵   [cyclic structure with CH₂, CH, CH₂, CH, CH₂]

<u>product</u>   <u>reactant</u>

9.70   [cyclohexane]$-CH=CH_2$

9.71 CH_3-[cyclopentene]$-CH_3$

9.72   (a) [structure with CH₃, Cl, Cl]   (b) [structures] + [structure]

$CH_3\overset{}{C}ClCH_2Cl$ [structure] $Cl\overset{}{C}(CH_3)_2$

In (b), the following intermediate carbocations are formed:

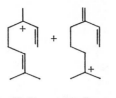

allylic   tertiary

Although the allylic carbocation is resonance-stabilized, the resonance structure shown is the major contributor because it is tertiary.

9.73   (a) 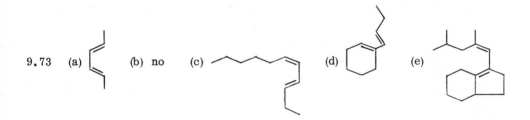   (b) no   (c)   (d)   (e)

9.74   All are 4n + 2. In (b), only the circled pi system would be involved in an electro-cyclic reaction.

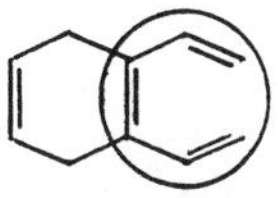

9.75   (a) conrotatory (4n, thermal)   (b) conrotatory (4n + 2, photo)
       (c) disrotatory (4n, photo)   (d) disrotatory (4n + 2, thermal)

9.76   (a)

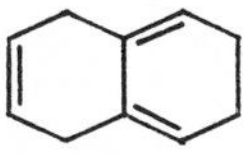

       (b) The cyclic reactant yields a 4n diene. The reaction is thermal and thus con-rotatory.

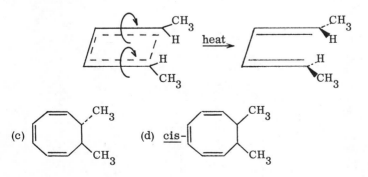

       (c)   (d) cis-

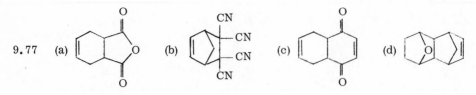

9.77   (a)   (b)   (c)   (d)

(e) A triple bond, rather than a double bond, is the reactive site of the dienophile. The product still contains a double bond at this position.

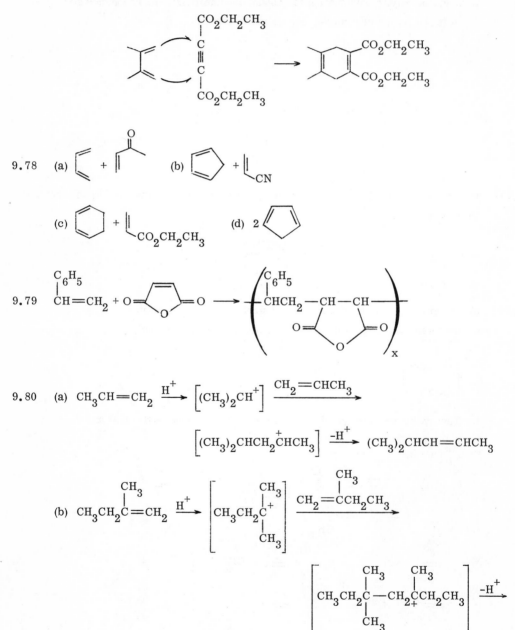

9.78   (a)   +   (b)   +

(c)   +   (d) 2

9.79 $CH=CH_2$ + ... → ...

9.80   (a) $CH_3CH=CH_2$ $\xrightarrow{H^+}$ $[(CH_3)_2CH^+]$ $\xrightarrow{CH_2=CHCH_3}$

$[(CH_3)_2CHCH_2\overset{+}{C}HCH_3]$ $\xrightarrow{-H^+}$ $(CH_3)_2CHCH=CHCH_3$

(b) $CH_3CH_2\underset{CH_3}{\overset{CH_3}{C}}=CH_2$ $\xrightarrow{H^+}$ $[CH_3CH_2\underset{CH_3}{\overset{CH_3}{\overset{+}{C}}}]$ $\xrightarrow{CH_2=CCH_2CH_3}$

$\left[CH_3CH_2\underset{CH_3}{\overset{CH_3}{C}}-CH_2\underset{2+}{\overset{CH_3}{C}}CH_2CH_3\right]$ $\xrightarrow{-H^+}$

$$CH_3CH_2\underset{\underset{CH_3}{|}}{\overset{\overset{CH_3}{|}}{C}}CH=\underset{}{\overset{\overset{CH_3}{|}}{C}}CH_2CH_3 + CH_3CH_2\underset{\underset{CH_3}{|}}{\overset{\overset{CH_3}{|}}{C}}CH_2\overset{\overset{CH_3}{|}}{C}=CHCH_3$$

9.81 (a) $CH_3CH_2CH_3 \xrightarrow{\underset{h\nu}{Br_2}} CH_3CHBrCH_3 \xrightarrow[\text{heat}]{OH^-} CH_3CH=CH_2$

(b) $CH_3CH_2CH_2Br \xrightarrow[\text{heat}]{KOC(CH_3)_3} CH_3CH=CH_2 \xrightarrow{Br_2} CH_3CHBrCH_2Br$

(c) $CH_3CH_2CH_2OH \xrightarrow[\text{heat}]{H_2SO_4} CH_3CH=CH_2 \xrightarrow{HBr} CH_3CHBrCH_3$

(d) [bicyclic structure] $\xrightarrow{NBS}$ (e) [cyclopentene] $\xrightarrow[\text{(2) Zn, HCl, H}_2\text{O}]{\text{(1) O}_3}$

(f) [cyclopentene] $\xrightarrow{H_2O, H^+}$ [cyclopentane]—OH

9.82 (a) $CH_3CH=CH_2$ [from 9.81(a)] $\xrightarrow[\text{Ni}]{D_2} CH_3CHDCH_2D$

(b) $CH_3CH=CH_2$ [from 9.81(b)] $\xrightarrow{KMnO_4} CH_3CO_2H$

(c) $CH_3CH=CH_2$ [from 9.81(a)] $\xrightarrow[\text{(2) H}_2\text{O}_2]{\text{(1) BH}_3} CH_3CH_2CH_2OH$

(d) [cyclohexane] $\xrightarrow[h\nu]{Br_2}$ [cyclohexane]—Br $\xrightarrow[\text{heat}]{KOH, CH_3CH_2OH}$

[cyclohexene] $\xrightarrow{C_6H_5CO_3H}$ [epoxide] $\xrightarrow{H_2O, H^+}$ [cyclohexane with OH, OH]

(e) [cyclopentane]—Cl $\xrightarrow[\text{heat}]{KOH, CH_3CH_2OH}$ [cyclopentene] $\xrightarrow{Cl_2, H_2O}$ [cyclopentane with Cl, OH]

9.83 The products from <u>cis</u>-3-hexene are diastereomers of those from <u>trans</u>-3-hexene.

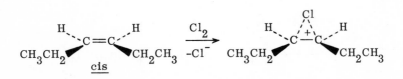

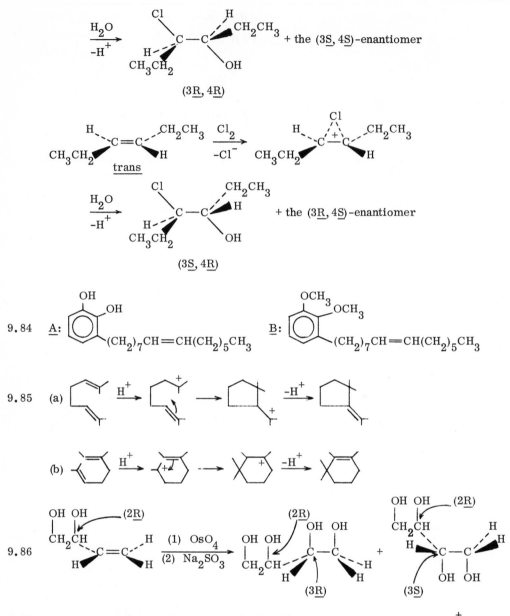

9.84 <u>A</u>: (structure with OH, OH, (CH$_2$)$_7$CH=CH(CH$_2$)$_5$CH$_3$) <u>B</u>: (structure with OCH$_3$, OCH$_3$, (CH$_2$)$_7$CH=CH(CH$_2$)$_5$CH$_3$)

9.85 (a)

(b)

9.86

(2R), (3R), (3S)

9.87 Treat 1-methyl-1-cyclopentene with the following reagents: (a) D$_2$O + H$^+$ (trace); (b) BD$_3$, then D$_2$O$_2$; (c) D$_2$, Ni catalyst; (d) DBr; (e) DBr + H$_2$O$_2$ (trace). These are all standard reactions that are usually performed with H$_2$O, BH$_3$, H$_2$, and HBr.

9.88 C$_{23}$H$_{46}$

H$_2$, Ni → C$_{23}$H$_{48}$

hot KMnO$_4$ → CH$_3$(CH$_2$)$_{12}$CO$_2$H + HO$_2$C(CH$_2$)$_7$CH$_3$

Br$_2$ → C$_{23}$H$_{46}$Br$_2$ (a pair of enantiomers)

The formula $C_{23}H_{46}$ (C_nH_{2n}) tells us that the structure contains one double bond or one ring. The H_2 reaction and the $KMnO_4$ reaction tell us that the structure contains one double bond. The $KMnO_4$ reaction shows the location of the double bond:

$$CH_3(CH_2)_{12}CH\!=\!CH(CH_2)_7CH_3$$

The fact that the Br_2 reaction yields a pair of enantiomers tells us that the housefly sex attractant is either cis-$CH_3(CH_2)_{12}CH\!=\!CH(CH_2)_7CH_3$ or the trans-isomer, and not a mixture of both. (A mixture would lead to two pairs of enantiomers.) It has been determined by other means that this alkene is the cis-compound.

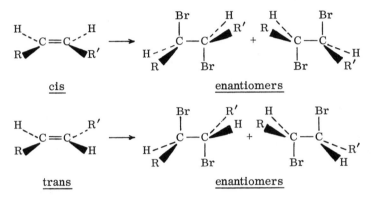

cis   enantiomers

trans   enantiomers

9.89   Each of the alkenes in (c) has the same number of alkyl substituents on the sp^2 carbon atoms. Also, in (c), one alkene contains an exocyclic double bond, and one contains an endocyclic double bond. The pair of alkenes in (c) is thus the best choice.

9.90   (a)

(b) The cis-isomer has a plane of symmetry and is superimposable on its mirror image; therefore, it does not exist as an enantiomer.

9.91   (a) and (b),

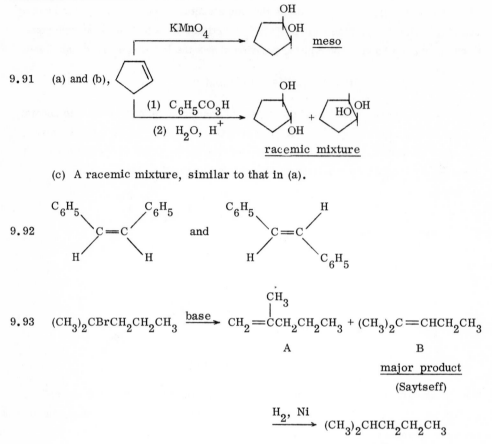

(c) A racemic mixture, similar to that in (a).

9.92

$$C_6H_5 \diagdown C = C \diagup C_6H_5$$
$$H \diagup \qquad \diagdown H$$

and

$$C_6H_5 \diagdown C = C \diagup H$$
$$H \diagup \qquad \diagdown C_6H_5$$

9.93 $(CH_3)_2CBrCH_2CH_2CH_3 \xrightarrow{\text{base}} CH_2{=}\overset{\overset{\displaystyle CH_3}{|}}{C}CH_2CH_2CH_3 + (CH_3)_2C{=}CHCH_2CH_3$

$$\qquad\qquad\qquad\qquad\qquad\qquad\qquad A \qquad\qquad\qquad\qquad\qquad B$$

$$\underline{\text{major product}}$$
$$\text{(Saytseff)}$$

$$\xrightarrow{H_2,\ Ni} (CH_3)_2CHCH_2CH_2CH_3$$

$$\text{2-methylpentane}$$

Chapter 10

Aromaticity, Benzene, and Substituted Benzenes

Some Important Features

An aromatic compound is one that is substantially stabilized by pi-electron delocalization. To be aromatic, a compound must be cyclic and planar; each ring atom must have a p orbital perpendicular to the plane of the ring; and the number of pi electrons in the ring must fit the formula $4n + 2$, where n is an integer.

Benzene, a typical aromatic compound, undergoes a variety of electrophilic aromatic substitution reactions.

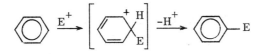

If the first substituent can donate electrons to the ring (by resonance or by the inductive effect), a second substitution occurs ortho and para because of stabilization of the intermediate (see Section 10.10).

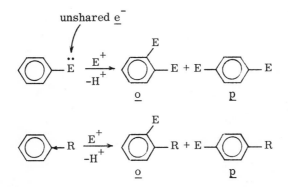

Except for the halogens, the o,p-directors activate the ring toward further electrophilic substitution.

If a first substituent cannot donate electronic charge to the ring, it deactivates the ring toward further electrophilic substitution. In this case, a second substitution occurs meta to the first substituent.

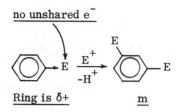

Ring is δ+   m

A ring substituted with an electron-withdrawing group is activated toward nucleo-philic substitution.

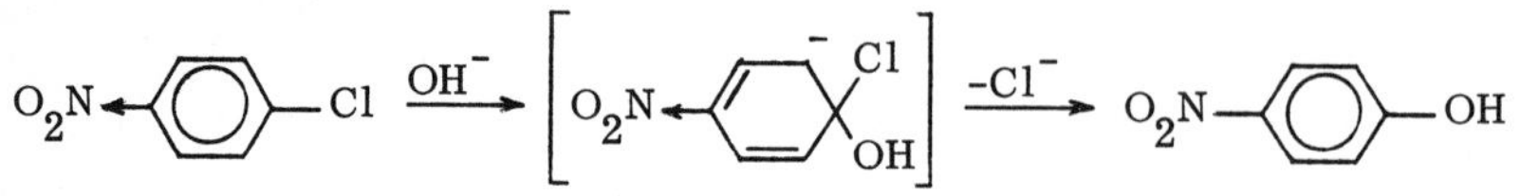

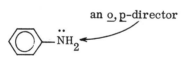

Aromatic nucleophilic substitutions can also occur under "forcing" conditions through a benzyne intermediate (Section 10.14).

The benzylic position of an alkylbenzene has enhanced reactivity toward oxidation and free-radical reactions (Section 10.13). Recall from Chapter 5 that benzyl halides are highly reactive in S_N1 and S_N2 reactions.

Reminders

Except for alkyl and aryl groups, an o, p-director has unshared electrons on the atom adjacent to the ring.

an o, p-director

Release of electron density by a substituent activates a ring toward electrophilic substitution, while withdrawal of electrons deactivates a ring.

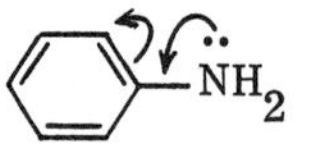

Ring is δ- and   Ring is δ+ and
activated toward E$^+$ deactivated toward E$^+$

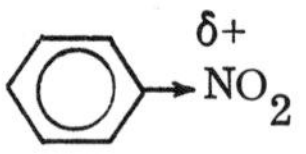

Answers to Problems

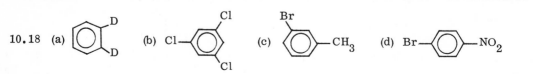

10.18 (a)   (b) Cl—   (c)   (d) Br—◯—NO_2

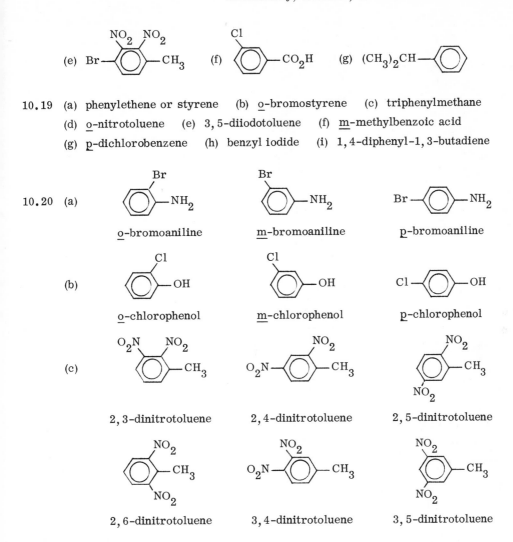

(e) Br—⟨benzene ring with NO$_2$, NO$_2$⟩—CH$_3$ (f) Cl—⟨benzene ring⟩—CO$_2$H   (g) (CH$_3$)$_2$CH—⟨benzene ring⟩

10.19 (a) phenylethene or styrene   (b) o-bromostyrene   (c) triphenylmethane
      (d) o-nitrotoluene   (e) 3,5-diiodotoluene   (f) m-methylbenzoic acid
      (g) p-dichlorobenzene   (h) benzyl iodide   (i) 1,4-diphenyl-1,3-butadiene

10.20 (a)

o-bromoaniline   m-bromoaniline   p-bromoaniline

(b)

o-chlorophenol   m-chlorophenol   p-chlorophenol

(c)

2,3-dinitrotoluene   2,4-dinitrotoluene   2,5-dinitrotoluene

2,6-dinitrotoluene   3,4-dinitrotoluene   3,5-dinitrotoluene

10.21 All the examples shown are aromatic. Compound (a) has 14 pi electrons. Compound (b) has ten. Compound (c) has ten. Compound (d) has 18. Compound (e) has 14.

The ten pi electrons in (c):

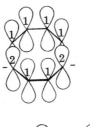

The 14 pi electrons in (e):

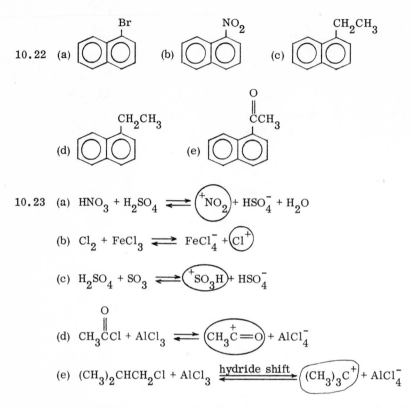

10.22 (a) [Br-substituted naphthalene] (b) [NO₂-substituted naphthalene] (c) [CH₂CH₃-substituted naphthalene]

(d) [CH₂CH₃-substituted naphthalene] (e) [COCH₃-substituted naphthalene]

10.23 (a) $HNO_3 + H_2SO_4 \rightleftharpoons \overset{+}{N}O_2 + HSO_4^- + H_2O$

(b) $Cl_2 + FeCl_3 \rightleftharpoons FeCl_4^- + \overset{+}{Cl}$

(c) $H_2SO_4 + SO_3 \rightleftharpoons \overset{+}{S}O_3H + HSO_4^-$

(d) $CH_3\overset{O}{\overset{\|}{C}}Cl + AlCl_3 \rightleftharpoons CH_3\overset{+}{C}{=}O + AlCl_4^-$

(e) $(CH_3)_2CHCH_2Cl + AlCl_3 \xrightarrow{\text{hydride shift}} (CH_3)_3\overset{+}{C} + AlCl_4^-$

10.24 The rate of reaction for nitrobenzene is much less than the rate for benzene because of electron-withdrawal by the nitro group.

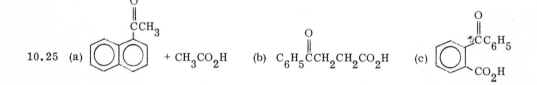

10.25 (a) [naphthalene with COCH₃ group] $+ CH_3CO_2H$ (b) $C_6H_5\overset{O}{\overset{\|}{C}}CH_2CH_2CO_2H$ (c) [benzene ring with COC₆H₅ and CO₂H groups]

10.26 (a) acetanilide, because the N has unshared electrons that help stabilize the intermediate by resonance.

(b) toluene, because the CH_3 group is electron-releasing by the inductive effect.

(c) p-xylene, because it has two electron-releasing groups on the ring.

(d) m-nitrotoluene, because it has an electron-releasing CH_3 group.

(e) chlorobenzene, because it has only one electron-withdrawing group.

(f) phenol, because the OH group has unshared electrons that stabilize the intermediate to a greater extent than the CH_3 group.

(g) phenol, for a reason similar to that in (f).

10.27 (a) o,p   (b) m   (c) o,p   (d) m   (e) m   (f) o,p   (g) o,p   (h) m
(i) o,p

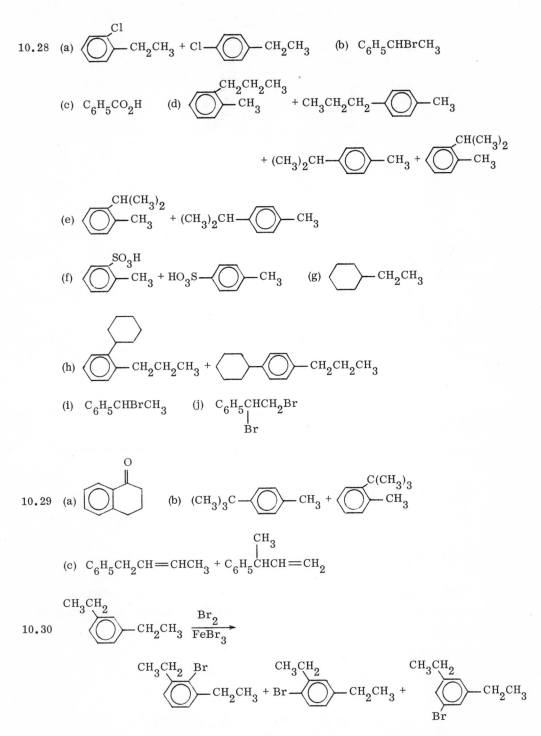

10.28 (a) [o-ethylchlorobenzene] + Cl—[benzene]—CH₂CH₃ (b) C₆H₅CHBrCH₃

(c) C₆H₅CO₂H (d) [structure] + CH₃CH₂CH₂—[benzene]—CH₃

+ (CH₃)₂CH—[benzene]—CH₃ + [structure]

(e) [structure] + (CH₃)₂CH—[benzene]—CH₃

(f) [structure] + HO₃S—[benzene]—CH₃ (g) [cyclohexane]—CH₂CH₃

(h) [structure] CH₂CH₂CH₃ + [structure]—CH₂CH₂CH₃

(i) C₆H₅CHBrCH₃ (j) C₆H₅CHCH₂Br with Br

10.29 (a) [tetralone structure] (b) (CH₃)₃C—[benzene]—CH₃ + [structure]

(c) C₆H₅CH₂CH=CHCH₃ + C₆H₅CHCH=CH₂ with CH₃

10.30 [structure] —CH₂CH₃ →(Br₂/FeBr₃)

[structures] —CH₂CH₃ + Br—[benzene]—CH₂CH₃ + [structure]—CH₂CH₃

By contrast, o-diethylbenzene yields two monobromo products, and p-diethyl-benzene yields only one.

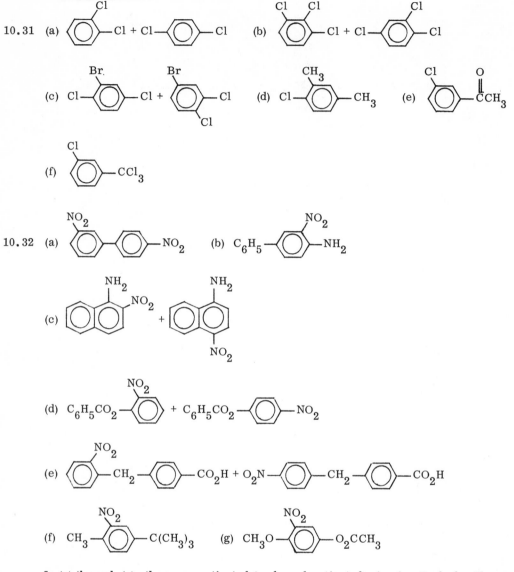

10.31 (a), (b), (c), (d), (e), (f)

10.32 (a), (b), (c), (d), (e), (f), (g)

In (a) through (e), the more activated (or less deactivated) ring is attacked. The position of substitution on this ring is determined by the substituent already on the ring. For example,

(d)

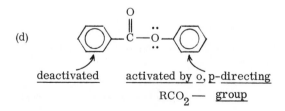

deactivated   activated by o, p-directing

RCO_2 —   group

If (f), attack occurs o to the CH_3 group (less hindered position). In (g), attack occurs o to the more strongly directing CH_3O group.

10.33 (a) $C_6H_5OCH_2CH_3$ $\xrightarrow{HNO_3}$

(b) $C_6H_5CH_3$ $\xrightarrow[AlCl_3]{CH_3Cl}$ $CH_3-\bigcirc-CH_3$ $\xrightarrow[\text{heat}]{KMnO_4}$

(c) $C_6H_5CH_3$ $\xrightarrow{HNO_3}$ $O_2N-\bigcirc-CH_3$ $\xrightarrow{Br_2}{FeBr_3}$

$O_2N-\overset{\overset{\displaystyle Br}{|}}{\bigcirc}-CH_3$ $\xrightarrow[\text{(2)\ OH}^-]{\text{(1)\ Fe,\ HCl}}$

10.34 (a) C_6H_6 $\xrightarrow[AlCl_3]{CH_3Cl}$ $C_6H_5CH_3$ $\xrightarrow[FeCl_3]{Cl_2}$ $Cl-\bigcirc-CH_3$ $\xrightarrow[\text{heat}]{KMnO_4}$

$Cl-\bigcirc-CO_2H$

(b) $[C_6H_5CH_3$ from (a)$]$ $\xrightarrow{HNO_3}$ $\overset{\overset{\displaystyle NO_2}{|}}{\bigcirc}-CH_3$

(c) excess C_6H_6 $\xrightarrow[AlCl_3]{CH_3CHCl_2}$ $CH_3CH(C_6H_5)_2$

(d) C_6H_6 $\xrightarrow[FeBr_3]{Br_2}$ C_6H_5Br $\xrightarrow[SO_3]{H_2SO_4}$ $Br-\bigcirc-SO_3H$

(e) $[C_6H_5CH_3$ from (a)$]$ $\xrightarrow[\text{heat}]{KMnO_4}$ $C_6H_5CO_2H$ $\xrightarrow[FeBr_3]{Br_2}$ $\overset{\overset{\displaystyle Br}{|}}{\bigcirc}-CO_2H$

(f) $[C_6H_5Br$ from (d)$]$ $\xrightarrow[H_2SO_4]{HNO_3}$ $Br-\bigcirc-NO_2$

(g) C_6H_6 $\xrightarrow[SO_3]{H_2SO_4}$ $C_6H_5SO_3H$ $\xrightarrow[\text{(2)\ H}_2\text{O,\ H}^+]{\text{(1)\ fuse with NaOH}}$ C_6H_5OH

10.35 $(C_6H_5)_3C^+ + H_2O + HSO_4^-$. The triphenylmethyl cation is highly resonance-stabilized because the positive charge is delocalized by the three benzene rings.

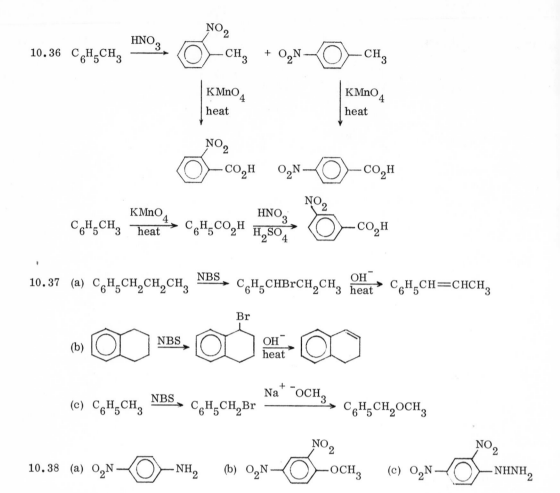

10.36 $C_6H_5CH_3$

10.37 (a) $C_6H_5CH_2CH_2CH_3 \xrightarrow{NBS} C_6H_5CHBrCH_2CH_3 \xrightarrow[heat]{OH^-} C_6H_5CH=CHCH_3$

(b)

(c) $C_6H_5CH_3 \xrightarrow{NBS} C_6H_5CH_2Br \xrightarrow{Na^+ \ ^-OCH_3} C_6H_5CH_2OCH_3$

10.38 (a) $O_2N-\!\!\!\bigcirc\!\!\!-NH_2$ (b) $O_2N-\!\!\!\bigcirc\!\!\!-OCH_3$ (c) $O_2N-\!\!\!\bigcirc\!\!\!-NHNH_2$

10.39   In effect, the reagent ICl must be split by the catalyst into a positive ion and a
negative ion. Because Cl is more electronegative than I, Cl is the negative end of
the reagent and I is the positive end. Therefore, I^+ is the attacking electrophile,
and the product is C_6H_5I.

10.40 (a)

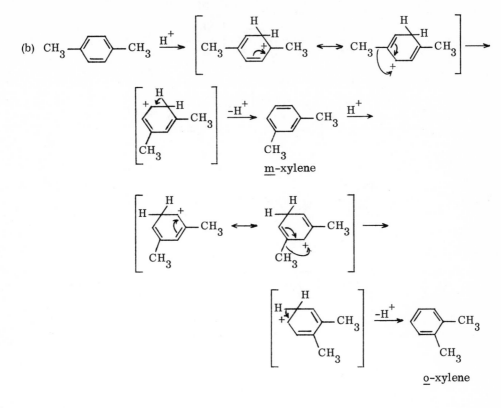

10.41 $C_6H_5C(CH_3)_3$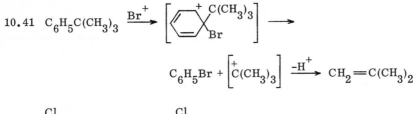

$C_6H_5Br + \left[\overset{+}{C}(CH_3)_3\right] \xrightarrow{-H^+} CH_2{=}C(CH_3)_2$

10.42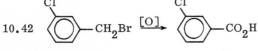

Because Br is lost in the oxidation, it cannot be attached to the ring.

10.43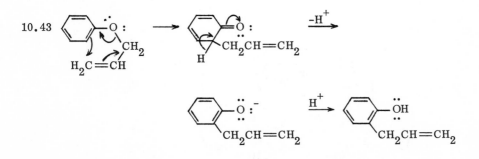

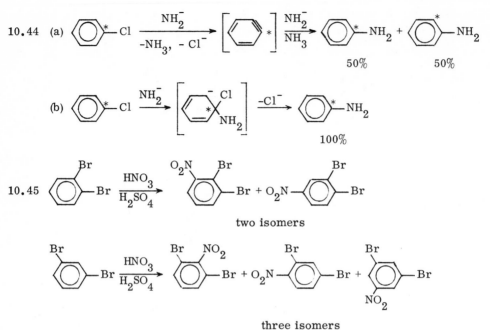

10.44 (a) and (b) schemes

10.45

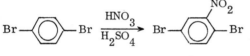

two isomers

three isomers

one isomer

The melting points of the three isomers are <u>ortho</u>, 6°; <u>meta</u>,−7°; <u>para</u>, 87°.

10.46 $(CH_3)_2N$—⬡—$\overset{\overset{O}{\|}}{C}$—⬡—$N(CH_3)_2$

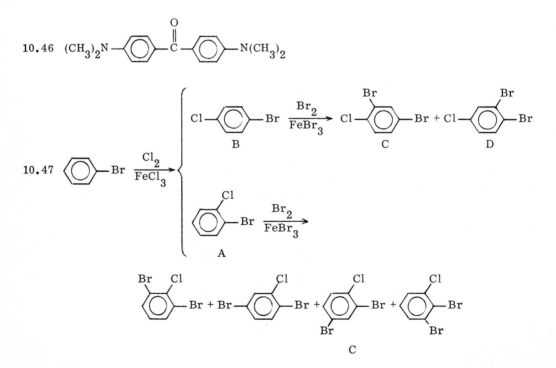

10.48 The unshared electrons on the nitrogen of aniline are delocalized and are less available for donation than those of cyclohexylamine. The positive charge in the product anilinium ion, by contrast, is not delocalized.

aniline:

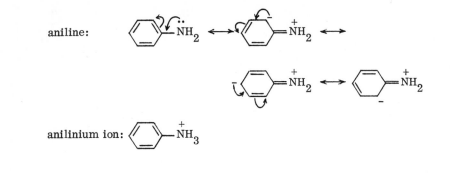

anilinium ion:

10.49 Less basic, because the nitro group helps delocalize the unshared electrons by resonance and by the inductive effect.

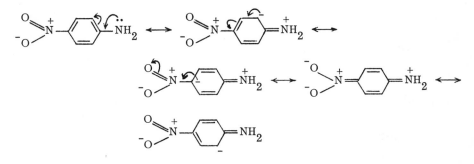

10.50 ortho:

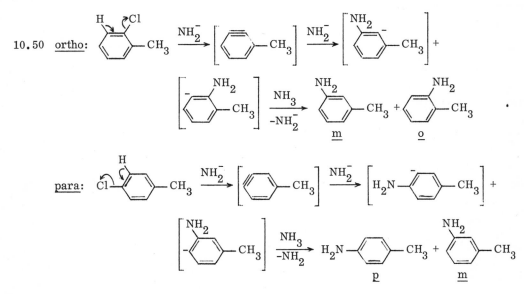

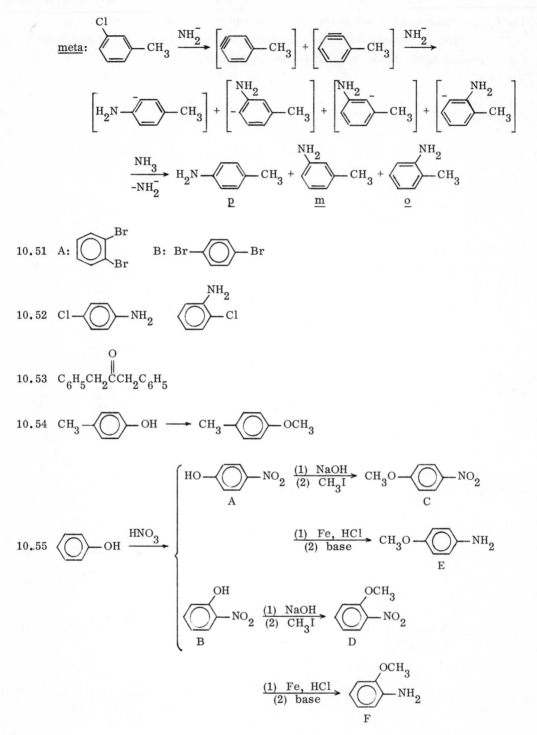

10.51   A: B:

10.52

O
‖
10.53   C₆H₅CH₂CCH₂C₆H₅

10.54

10.55

Chapter 11

Aldehydes and Ketones

Some Important Features

A carbonyl group (C=O) is polar and can be attacked by an electrophile such as H$^+$ or by a nucleophile.

<u>In acidic solution, the oxygen is protonated:</u>

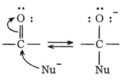

<u>resonance structures for the</u>
<u>protonated carbonyl group</u>

<u>In base, the carbon is attacked:</u>

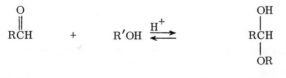

Aldehydes and ketones can undergo many <u>addition reactions</u> (see the summary to Chapter 11). Many of these reactions are reversible; some, such as Grignard reactions, are irreversible.

$$
\underset{\text{an aldehyde}}{\overset{\overset{\displaystyle O}{\|}}{R\overset{}{C}H}} \quad + \quad R'OH \underset{}{\overset{H^+}{\rightleftarrows}} \quad \underset{\text{a hemiacetal}}{\overset{\overset{\displaystyle OH}{|}}{\underset{\underset{\displaystyle OR}{|}}{R\overset{}{C}H}}}
$$

<table>
<tr><td align="center"><u>favored by a less positive,
more hindered C=O carbon,
as in a ketone</u></td><td align="center"><u>favored by a more positive,
less hindered C=O carbon,
as in an aldehyde</u></td></tr>
</table>

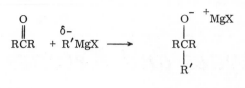

a ketone   a magnesium alkoxide

Aldehydes and ketones can undergo <u>addition-elimination reactions</u> with many nitrogen compounds. The products are favored by resonance-stabilization or by electron-withdrawing groups on the nitrogen. Typical reagents are primary amines (RNH_2) and hydrazine derivatives ($RNHNH_2$).

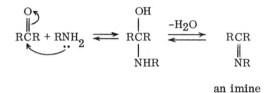

<u>an imine</u>

Aldehydes and ketones undergo catalytic hydrogenation with heat and pressure; however, these conditions also reduce carbon-carbon double bonds. Reduction to an alcohol with a metal hydride does not usually reduce a carbon-carbon double bond. Aldehydes are also very easily oxidized to carboxylic acids.

A hydrogen atom α to a carbonyl group is slightly acidic. (For resonance structures of the product enolate ions, see Section 11.16.)

$$R_2C\overset{O}{\overset{\|}{C}}R \ +\ ^-OR \ \rightleftharpoons\ R_2\bar{C}\overset{O}{\overset{\|}{C}}R \ +\ HOR$$
(with H on the carbon)

$$RC\overset{O}{\overset{\|}{C}}CH\overset{O}{\overset{\|}{C}}R \ +\ ^-OR \ \rightleftharpoons\ RC\overset{O}{\overset{\|}{C}}\bar{C}H\overset{O}{\overset{\|}{C}}R \ +\ HOR$$
(with H on the carbon)

An aldehyde or ketone with an α hydrogen can undergo <u>tautomerism.</u> Unless the enol tautomer is stabilized relative to the keto tautomer (by hydrogen bonding, for example), the keto form is favored.

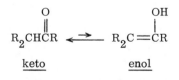

keto   enol

Aldehydes or ketones with the carbonyl group in conjugation with a carbon-carbon double bond can undergo <u>1,4-addition reactions</u> (either electrophilic or nucleophilic). An isolated carbon-carbon double bond cannot participate in a 1,4-addition reaction.

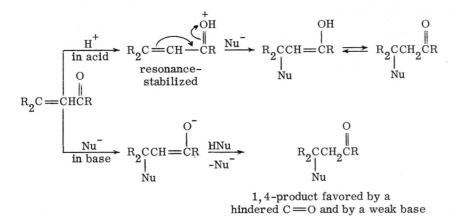

1,4-product favored by a
hindered C=O and by a weak base

Other important topics covered in this chapter are the synthesis of enamines (Section 11.10B), the Wittig reaction for alkenes (Section 11.12), and α halogenation (Section 11.18).

Reminders

In acidic solution, the C=O oxygen is protonated, and a weak nucleophile can attack the C=O carbon.

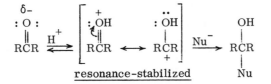

A weak base, such as ⁻CN, attacks the C=O carbon.

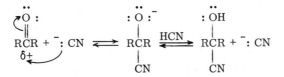

A strong base, such as ⁻OR, can remove an α hydrogen.

Remember that an sp^2 carbon atom is, of itself, not a chiral center in a molecule.

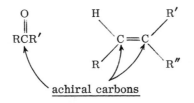

achiral carbons

Answers to Problems

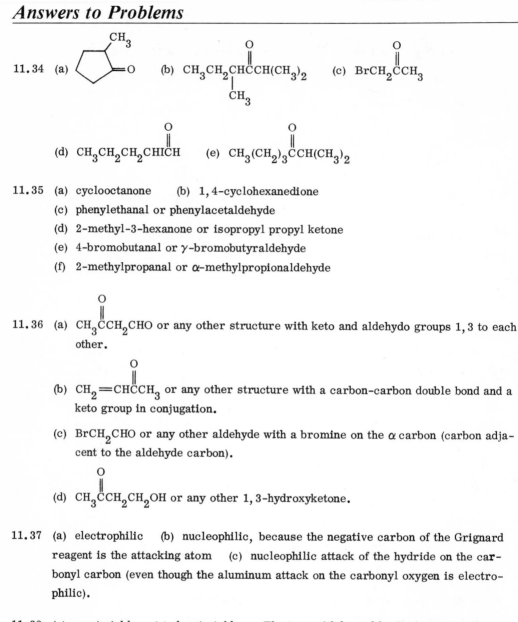

11.34 (a) [cyclopentane ring with CH$_3$ substituent and =O]   (b) $CH_3CH_2CHCCH(CH_3)_2$ with carbonyl O and CH$_3$ branch   (c) $BrCH_2CCH_3$ with carbonyl O

(d) $CH_3CH_2CH_2CHICH$ with carbonyl O   (e) $CH_3(CH_2)_3CCH(CH_3)_2$ with carbonyl O

11.35 (a) cyclooctanone   (b) 1,4-cyclohexanedione
(c) phenylethanal or phenylacetaldehyde
(d) 2-methyl-3-hexanone or isopropyl propyl ketone
(e) 4-bromobutanal or γ-bromobutyraldehyde
(f) 2-methylpropanal or α-methylpropionaldehyde

11.36 (a) CH_3CCH_2CHO (with carbonyl O) or any other structure with keto and aldehydo groups 1,3 to each other.

(b) $CH_2{=}CHCCH_3$ (with carbonyl O) or any other structure with a carbon–carbon double bond and a keto group in conjugation.

(c) $BrCH_2CHO$ or any other aldehyde with a bromine on the α carbon (carbon adjacent to the aldehyde carbon).

(d) $CH_3CCH_2CH_2OH$ (with carbonyl O) or any other 1,3-hydroxyketone.

11.37 (a) electrophilic   (b) nucleophilic, because the negative carbon of the Grignard reagent is the attacking atom   (c) nucleophilic attack of the hydride on the carbonyl carbon (even though the aluminum attack on the carbonyl oxygen is electrophilic).

11.38 (c) most stable   (a) least stable.   Electron withdrawal by Br increases the positive charge on the carbonyl carbon; three bromine atoms are more effective than one.

11.39 $Br_2CHCHO + H_2O \rightleftharpoons Br_2CHCH(OH)_2$

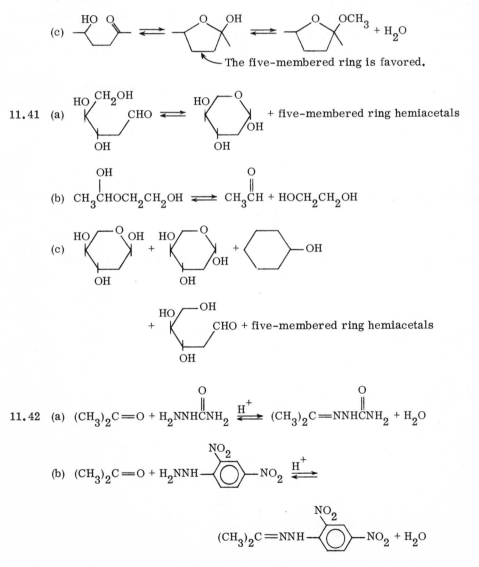

11.40 (a) $CH_3CH_2CHO + CH_3OH \rightleftharpoons CH_3CH_2\overset{\displaystyle OH}{\underset{|}{C}}HOCH_3 \overset{CH_3OH}{\rightleftharpoons}$

$$CH_3CH_2CH(OCH_3)_2 + H_2O$$

(b) $(CH_3)_2C{=}O + HOCH_2CH_2OH \rightleftharpoons (CH_3)_2\overset{\displaystyle OH}{\underset{|}{C}}OCH_2CH_2OH \rightleftharpoons$

$$(CH_3)_2C\underset{O}{\overset{O}{\diagdown}}\!\!\Big] + H_2O$$

Since a five-membered ring can be formed in the hemiacetal-to-acetal step, the
cyclization is favored over reaction with another molecule of ethylene glycol.

(c)

— The five-membered ring is favored.

11.41 (a)

+ five-membered ring hemiacetals

(b) $CH_3\overset{\displaystyle OH}{\underset{|}{C}}HOCH_2CH_2OH \rightleftharpoons CH_3\overset{\displaystyle O}{\overset{\|}{C}}H + HOCH_2CH_2OH$

(c)

$+$ $+$ $-OH$

+   CHO + five-membered ring hemiacetals

11.42 (a) $(CH_3)_2C{=}O + H_2NNH\overset{\displaystyle O}{\overset{\|}{C}}NH_2 \overset{H^+}{\rightleftharpoons} (CH_3)_2C{=}NNH\overset{\displaystyle O}{\overset{\|}{C}}NH_2 + H_2O$

(b) $(CH_3)_2C{=}O + H_2NNH{-}\!\!\underset{NO_2}{\overset{NO_2}{\bigcirc}}\!\!{-}NO_2 \overset{H^+}{\rightleftharpoons}$

$$(CH_3)_2C{=}NNH{-}\!\!\underset{NO_2}{\overset{NO_2}{\bigcirc}}\!\!{-}NO_2 + H_2O$$

11.43 (a) $C_6H_5CHO + H_2NC_6H_5 \xrightarrow{H^+} C_6H_5CH{=}NC_6H_5 + H_2O$

(b) $CH_3CH_2CHO + HN(CH_3)_2 \xrightarrow{H^+} CH_3CH{=}CHN(CH_3)_2 + H_2O$

11.44 (a) $CH_3CH_2CH_2CHO + HN(CH_2CH_3)_2 \xrightarrow{H^+}$

(b)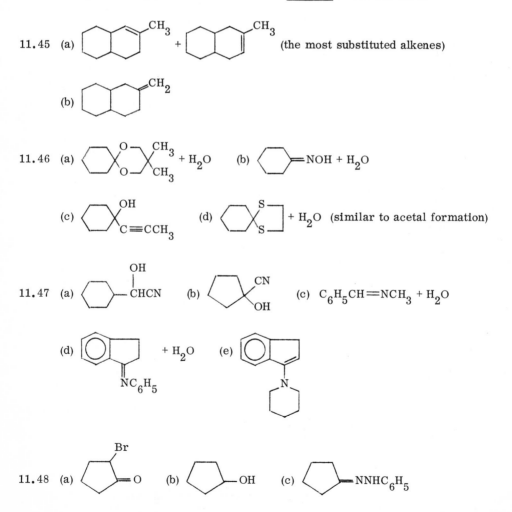
=O + HN $\xrightarrow{H^+}$

(c) $CH_3CH_2CHO + C_6H_5NH_2 \xrightleftharpoons{-H_2O}$

The products of (a) and (b) are enamines because <u>secondary</u> amines were used.
The product of (c) is an imine because a <u>primary</u> amine was used.

11.45 (a) + (the most substituted alkenes)

(b)

11.46 (a) + H_2O (b) =NOH + H_2O

(c)   (d) + H_2O (similar to acetal formation)

11.47 (a)   (b)   (c) $C_6H_5CH{=}NCH_3 + H_2O$

(d) + H_2O (e)

11.48 (a) =O   (b) —OH   (c) =NNHC$_6$H$_5$

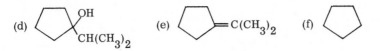

(d) (e) (f)

11.49 (b), (a), (c). Compound (b) contains the most hindered carbonyl group, and compound (c) contains the least hindered carbonyl group.

11.50 The structure of the required Wittig reagent may be determined by circling the portion of the product that has replaced the carbonyl oxygen.

Example:

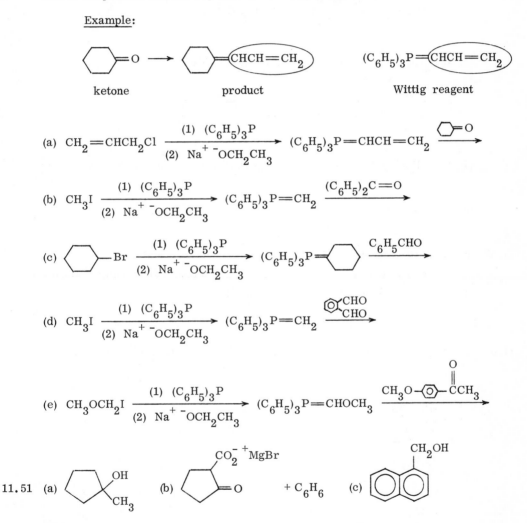

ketone   product   Wittig reagent

(a) CH_2=$CHCH_2Cl$ $\xrightarrow[\text{(2) } Na^+ \text{ }^-OCH_2CH_3]{\text{(1) } (C_6H_5)_3P}$ $(C_6H_5)_3P$=$CHCH$=CH_2 $\longrightarrow$

(b) CH_3I $\xrightarrow[\text{(2) } Na^+ \text{ }^-OCH_2CH_3]{\text{(1) } (C_6H_5)_3P}$ $(C_6H_5)_3P$=CH_2 $\xrightarrow{(C_6H_5)_2C=O}$

(c) Br $\xrightarrow[\text{(2) } Na^+ \text{ }^-OCH_2CH_3]{\text{(1) } (C_6H_5)_3P}$ $(C_6H_5)_3P$= $\xrightarrow{C_6H_5CHO}$

(d) CH_3I $\xrightarrow[\text{(2) } Na^+ \text{ }^-OCH_2CH_3]{\text{(1) } (C_6H_5)_3P}$ $(C_6H_5)_3P$=CH_2 $\xrightarrow{\substack{CHO \\ CHO}}$

(e) CH_3OCH_2I $\xrightarrow[\text{(2) } Na^+ \text{ }^-OCH_2CH_3]{\text{(1) } (C_6H_5)_3P}$ $(C_6H_5)_3P$=$CHOCH_3$ $\xrightarrow{CH_3O-\bigcirc-\overset{O}{\overset{\|}{C}}CH_3}$

11.51 (a) OH CH$_3$ (b) =O CO_2^- ^+MgBr + C_6H_6 (c) CH_2OH

In (b), the starting carbonyl compound contains an acidic proton that reacts with the Grignard reagent at a faster rate than does the keto group.

11.52 $(CH_3)_2CHOH$ $\xrightarrow{KMnO_4}$ $(CH_3)_2C$=O $\left.\begin{array}{c} \\ \\ \end{array}\right\}$ $\xrightarrow{H_2O, \text{ } H^+}$ $(CH_3)_2\overset{OH}{\underset{|}{C}}C_6H_5$

C_6H_5Br $\xrightarrow{Mg}$ C_6H_5MgBr

11.53 (a) $CH_3I \xrightarrow{Mg} CH_3MgI \xrightarrow[\text{(2) } H_2O, \text{ } H^+]{\text{(1) } (CH_3CH_2)_2C=O} CH_3CH_2\overset{\overset{\displaystyle OH}{|}}{\underset{\underset{\displaystyle CH_3}{|}}{C}}CH_2CH_3$

   (b) $CH_3MgI \xrightarrow[\text{(2) } H_2O, \text{ } H^+]{\text{(1) } HCHO} CH_3CH_2OH$

   (c) $CH_3MgI \xrightarrow[\text{(2) } H_2O, \text{ } H^+]{\text{(1) } CH_3CH_2CH_2CHO} CH_3CH_2CH_2\overset{\overset{\displaystyle OH}{|}}{C}HCH_3$

11.54 (a) $(CH_3)_2CHNH_2$ (b) $(CH_3)_2CHNHCH_2CH_2OH$

   (c) by way of a cyclic imine,

   (d) $CH_3CH_2CH_2OH$ (e) $HOCH_2CH_2CH_2CH_2OH$ (f) no reaction

   (g) $-CO_2^-$ (h) $-CO_2^- + CHCl_3$

11.55 (a) and (c). Compound (c) is a hemiacetal, which is in equilibrium with the alde-
       hyde. Compound (d) is an acetal, which is not in equilibrium with the aldehyde
       under the alkaline conditions of the Tollens test.

11.56 (a) $CH_3CH_2C\underset{\underset{}{}}{H}_2CO_2CH_2CH_3$ (b)   (c) $CH_3\overset{\overset{\displaystyle O}{||}}{C}C\overset{H}{\underset{\underset{\displaystyle CH_3}{|}}{}}\overset{\overset{\displaystyle O}{||}}{C}CH_2CH_3$

   (d)

11.57 (b), (c), (a)

11.58 (a) $\rightleftharpoons {}^-CH(CO_2CH_2CH_3)_2 + HOCH_2CH_3$ (b) $\rightleftharpoons {}^-CH_2CHO + H_2O$

   (c) $\rightleftharpoons O=$$=O + HCO_3^-$

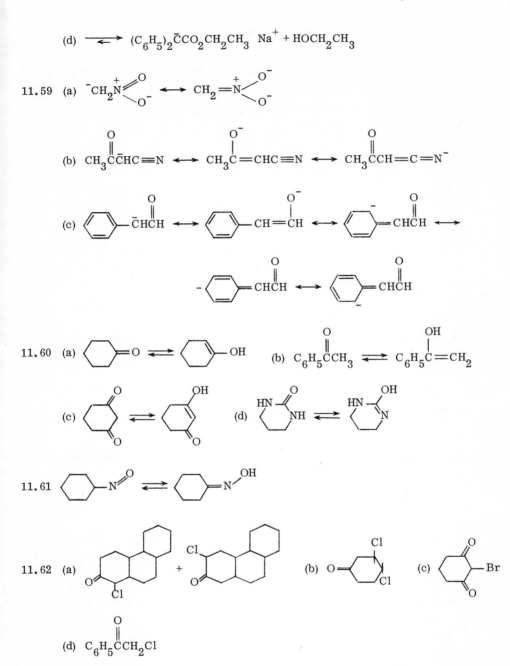

(d) $\rightleftharpoons$ $(C_6H_5)_2\bar{C}CO_2CH_2CH_3$ Na^+ + $HOCH_2CH_3$

11.59 (a)

(b)

(c)

11.60 (a)   (b)

(c)   (d)

11.61

11.62 (a)   + (b)   (c)

(d) $C_6H_5\overset{O}{\overset{\|}{C}}CH_2Cl$

11.63 (a), (d), and (e) would yield a yellow precipitate of CHI_3; (b), (c), and (f) would not show this precipitate because none is a methyl ketone (and none can be oxidized to such a structure).

11.64 (a) Add a solution of 2,4-dinitrophenylhydrazine; cyclohexanone gives a precipitate, while cyclohexanol does not.

(b) Add I_2 in dilute NaOH; 2-pentanone gives a yellow precipitate, while 3-penta-none does not.

(c) Add Tollens reagent; pentanal gives a silver mirror, while 2-pentanone does not.

(d) Add Br_2 in CCl_4; 2-pentene decolorizes the solution, while 2-pentanone does not.

11.65 (a) HCl   (b) NaCN   (c) $CH_3CH_2NH_2$

11.66 (a) $CH_3CHICH_2CO_2CH_2CH_3$ (b)

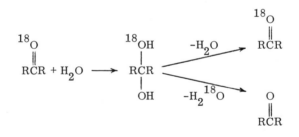

(c) $BrCH_2CH_2\overset{\displaystyle O}{\overset{\|}{C}}CH_3$ (d) $CH_3CHClCH_2CH_2\overset{\displaystyle O}{\overset{\|}{C}}CH_3$ (e)

In (d), the double bond is not in conjugation, and therefore Markovnikov's rule applies.

11.67 Measure the rate of ^{18}O-uptake by water; this rate is half the rate of hydration and subsequent dehydration.

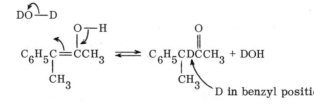

11.68 $C_6H_5\overset{\displaystyle OH}{\underset{\displaystyle CH_3}{\overset{\displaystyle |}{\underset{|}{C}}}}=CCH_3$ is the preferred enol because the double bond is in conjugation with the benzene ring. In acidic or basic solution, the D_2O reacts with this enol as follows:

$$DO\overset{\frown}{-}D$$

$$C_6H_5\underset{\displaystyle CH_3}{\overset{\displaystyle O\,-\,H}{\overset{|}{\underset{|}{C}}}}=CCH_3 \rightleftharpoons C_6H_5\underset{\displaystyle CH_3}{\overset{\displaystyle O}{\overset{\|}{\underset{|}{C}}}}DCCH_3 + DOH$$

D in benzyl position

11.69 The ketone undergoes racemization because the chiral center is α to the carbonyl group and undergoes tautomerism at that position.

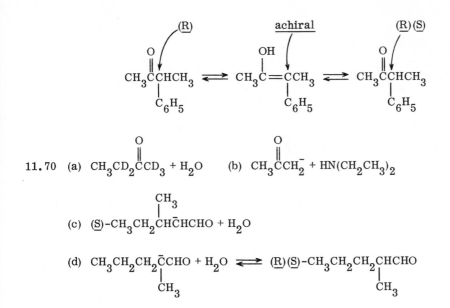

11.70 (a) $CH_3CD_2\overset{O}{\overset{\|}{C}}CD_3 + H_2O$ (b) $CH_3\overset{O}{\overset{\|}{C}}CH_2^- + HN(CH_2CH_3)_2$

(c) $(S)\text{-}CH_3CH_2\overset{\overset{CH_3}{|}}{CH}\overset{-}{C}HCHO + H_2O$

(d) $CH_3CH_2CH_2\overset{\overset{O}{\|}}{C}CHO + H_2O \rightleftharpoons (R)(S)\text{-}CH_3CH_2CH_2\overset{\overset{}{|}}{C}HCHO$
   $\overset{|}{CH_3}$ $\overset{|}{CH_3}$

In (c) and (d), note the position of the chiral carbon atom in reference to the carbonyl group.

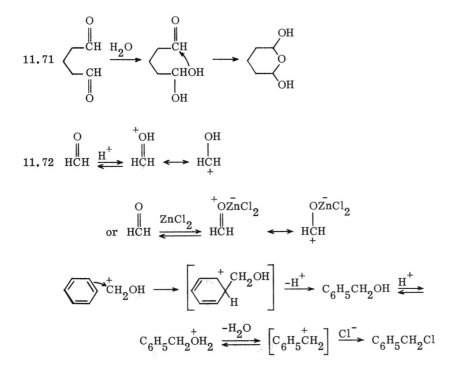

11.71

11.72

11.73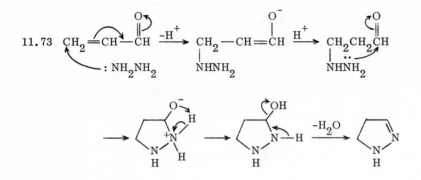

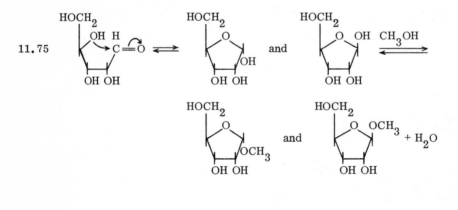

11.74 The aldehyde group must be converted to a functional group that is inert to a Grignard reagent. A cyclic acetal is a suitable blocking group for the aldehyde group.

$$BrCH_2CH_2CHO + HOCH_2CH_2OH \xrightarrow[-H_2O]{H^+}$$

$$BrCH_2CH_2\overset{O}{\underset{O}{CH}} \xrightarrow[\text{ether}]{Mg} BrMgCH_2CH_2\overset{O}{\underset{O}{CH}}$$

At this point, the desired Grignard reaction would be carried out, and then the
aldehyde would be regenerated in the hydrolysis step.

11.75

11.76

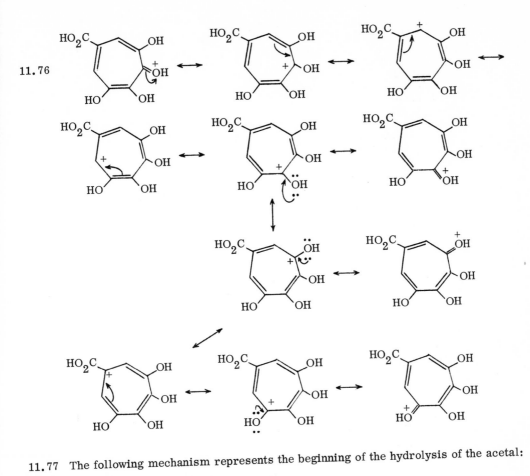

11.77   The following mechanism represents the beginning of the hydrolysis of the acetal:

$$\underset{CH_3CHOCH_2CH_3}{\overset{OCH_2CH_3}{|}} \xrightarrow{H^+} \underset{CH_3CHOCH_2CH_3}{\overset{\overset{+}{\underset{}{C}}OCH_2CH_3 \overset{H}{}}{}} \xrightarrow{-CH_3CH_2OH}$$

$$\left[CH_3\overset{+}{C}HOCH_2CH_3 \longleftrightarrow CH_3CH=\overset{+}{O}CH_2CH_3 \right]$$

$$\underline{\text{resonance-stabilized carbocation}}$$

The same steps would not occur for diethyl ether because a primary carbocation would be an intermediate.

$$CH_3CH_2OCH_2CH_3 \xrightarrow{H^+} \underset{\overset{|}{H}}{CH_3CH_2\overset{+}{O}CH_2CH_3} \xrightarrow{-CH_3CH_2OH} \left[CH_3\overset{+}{C}H_2 \right]$$

$$\underline{\text{not formed}}$$

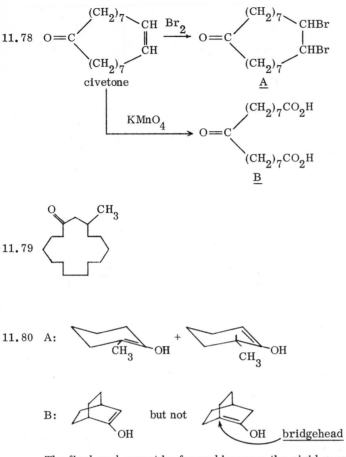

11.78

civetone → A

KMnO$_4$ → B

11.79

11.80 A:

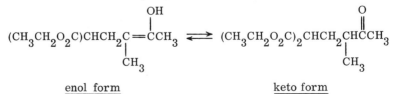

B: but not OH bridgehead

The final enol cannot be formed because the rigid geometry of the ring system does not allow p-orbital overlap between the bridgehead carbon and the carbonyl carbon.

11.81 $(CH_3CH_2O_2C)_2\bar{C}H$ $CH_2{=}C{-}\overset{\overset{O}{\|}}{C}CH_3$ $\overset{1,4 \text{ addition}}{\longrightarrow}$
CH$_3$

$(CH_3CH_2O_2C)_2CHCH_2C{=}\overset{\overset{O^-}{|}}{C}CH_3$ $\overset{H^+}{\longrightarrow}$
CH$_3$

$(CH_3CH_2O_2C)CHCH_2\overset{\overset{OH}{|}}{C}{=}CCH_3$ ⇌ $(CH_3CH_2O_2C)_2CHCH_2CH\overset{\overset{O}{\|}}{C}CH_3$
CH$_3$ CH$_3$

enol form   keto form

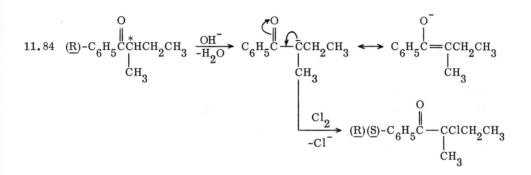

11.82

11.83 In acid, the hemiacetal OH can be protonated and forms a leaving group $\left(-OH_2^+\right)$.
       In base, no such leaving group can be formed.

11.84

11.85 I (d)   II (c)   III (b)

11.86 A: $C_6H_5CH{=}CHCHO$ B: $C_6H_5CH{=}CHCH_2OH$

Chapter 12

Carboxylic Acids

Some Important Features

Carboxylic acids (RCO_2H) are more acidic than alcohols, phenols, or carbonic acid. Carboxylic acids yield carboxylate salts when treated with any base stronger than the carboxylate ion itself.

$$RCO_2H + HCO_3^- \rightleftharpoons RCO_2^- + H_2O + CO_2$$

Factors affecting acid strength are discussed in detail in this chapter of the text. Among these factors are the <u>electronegativity</u> of the atom attached to the acidic hydrogen. A greater electronegativity means a stronger acid; therefore, alkanes are among the weakest acids known, while amines (R_2NH), alcohols (ROH), and hydrogen halides (HX) are successively stronger acids.

<u>Hybridization</u> of the atom attached to H affects acidity: $RC{\equiv}CH$ is more acidic than $R_2C{=}CHR$ or RH.

The <u>inductive effect</u> increases or decreases acid strength by withdrawing or releasing electron density.

$$ClCH_2CO_2H \rightleftharpoons ClCH_2CO_2^- + H^+$$

Cl strengthens the acid by stabilizing the anion through dispersal of the negative charge.

$$Cl{-}\langle\bigcirc\rangle{-}CO_2H \rightleftharpoons Cl{-}\langle\bigcirc\rangle{-}CO_2^- + H^+$$

<u>Resonance-stabilization</u> of the anion also increases acid strength. This anion stabilization is the primary reason for the acidity of carboxylic acids. Resonance-stabilization of the anion also explains why phenols are more acidic than alcohols.

$$RC\overset{O}{\overset{\|}{O}}H \rightleftharpoons \left[RC\overset{:\ddot{O}:}{\overset{\|}{{-}\ddot{O}:^-}} \longleftrightarrow RC\overset{:\ddot{O}:^-}{{=}\ddot{O}:} \right] + H^+$$

resonance structures

Other factors, such as hydrogen bonding in the anion, can also affect acid strength.

A carboxylic acid may undergo esterification with an alcohol. The rate of esterification decreases with increasing steric hindrance around the carboxyl and hydroxyl groups. (Review the mechanism for this reaction in Section 12.8.)

$$\underset{}{\overset{O}{\overset{\|}{R C}}OH} + R'OH \underset{}{\overset{H^+}{\rightleftharpoons}} \underset{\text{an ester}}{\overset{O}{\overset{\|}{R C}}OR'} + H_2O$$

<u>Decarboxylation</u> of β-keto acids is important both in the laboratory and in biological systems.

$$\underset{\text{a }\beta\text{-keto acid}}{\overset{O\quad O}{R\overset{\|}{C}CH_2\overset{\|}{C}OH}} \xrightarrow[\text{or enzymes}]{\text{heat}} \overset{O}{R\overset{\|}{C}CH_3} + CO_2$$

Other reactions of carboxylic acids are reduction by LAH (Section 12.9), 1,4-addition to α,β-unsaturated carboxylic acids (Section 12.10D), and anhydride formation by diacids (Section 12.10A and B).

Reminders

For calculations of K_a and pK_a values, review Section 1.10.

Remember that the anion of a strong acid is a weak base, while the anion of a weak acid is a strong base.

$$\underset{\text{strong acid}}{RCO_2H} \rightleftharpoons \underset{\text{weak base}}{RCO_2^-} + H^+$$

$$\underset{\text{weak acid}}{ROH} \rightleftharpoons \underset{\text{strong base}}{RO^-} + H^+$$

Because a carboxylic acid contains oxygen atoms with unshared electrons, it can be protonated by a strong acid or it can undergo hydrogen bonding.

$$\overset{\ddot{O}:}{R\overset{\|}{C}OH} \overset{H^+}{\rightleftharpoons} \left[\overset{+\ddot{O}H}{R\overset{\|}{C}OH} \longleftrightarrow \overset{:\ddot{O}H}{R\overset{|}{C}OH} \longleftrightarrow RC\overset{:\ddot{O}H}{=}\overset{}{OH} \right]$$

resonance structures for
protonated acid

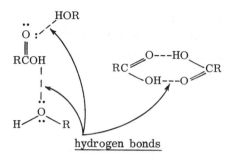

hydrogen bonds

Answers to Problems

12.22 (a) 2,2-dimethylpropanoic acid   (b) p-chlorobenzoic acid

(c) 2,3-dibromopentanoic acid or α,β-dibromovaleric acid

(d) magnesium propanoate   (e) calcium ethanedioate or calcium oxalate

(f) sodium 2-bromopropanoate or sodium α-bromopropionate

12.23 (a) $CH_2ICH_2CH_2CO_2H$ (b) $HCO_2^-\ K^+$ (c)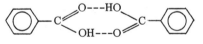

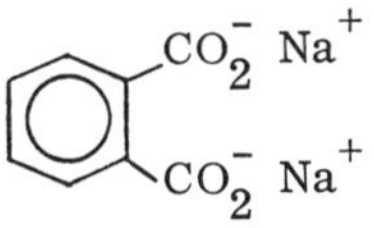

(d) $C_6H_5CO_2^-\ Na^+$ (e)

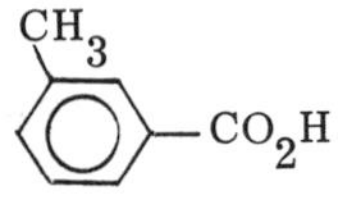

12.24 (a) $CH_3CH_2\overset{O}{\underset{\|}{C}}-$ (b) $CH_3CH_2CH_2\overset{O}{\underset{\|}{C}}-$ (c)

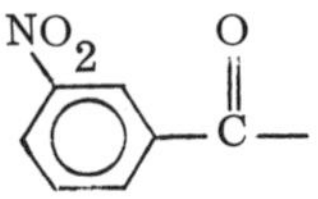

12.25 Benzoic acid undergoes hydrogen bonding; ethyl benzoate does not.

benzoic acid dimer

12.26 (a) H_2O---H_2O

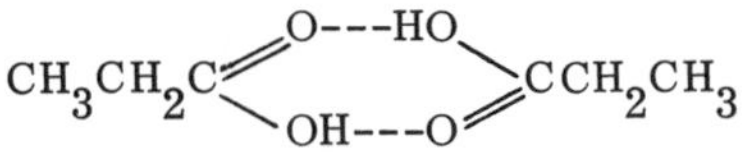

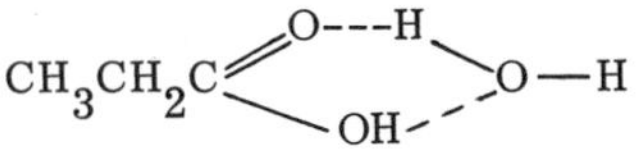

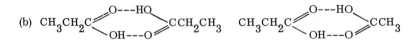

(b)

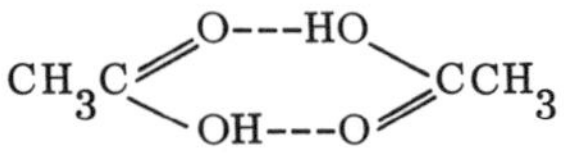

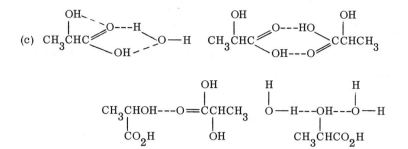

(c)

(There are other possibilities.)

12.27 (a) $CH_3CH_2CH_2Br$ $\xrightarrow{KCN}$ $CH_3CH_2CH_2CN$ $\xrightarrow[\text{heat}]{H_2O, H^+}$ $CH_3CH_2CH_2CO_2H$

$\searrow Mg$

$\nearrow H_2O, H^+$

$CH_3CH_2CH_2MgBr$ $\xrightarrow{CO_2}$ $CH_3CH_2CH_2CO_2MgBr$

(b) $CH_3CH_2CH_2CH_2OH$ $\xrightarrow{KMnO_4}$

(c) $CH_3CH_2CH_2CHO$ $\xrightarrow{KMnO_4}$

(d) $CH_3CH_2CH_2CH{=}CHCH_2CH_2CH_3$ $\xrightarrow[\text{heat}]{KMnO_4}$

In (b), (c), and (d), other oxidizing agents would also be suitable.

(e) $\xrightarrow[\text{heat}]{H_2O, H^+}$ (f) $\xrightarrow[\text{heat}]{H_2O, H^+}$

12.28 (a) (1) KCN   (2) dil. HCl, heat   (3) neutralize with NaOH

(b) (1) Mg, diethyl ether   (2) CO_2 (3) dil. HCl

(c) (1) CrO_3, pyridine   (2) HCN   (3) dil. HCl, heat

(d) (1) excess KCN   (2) dil. HCl, heat

(e) (1) HBr, H_2O_2 (2) KCN   (3) dil. HCl, heat

(f) (1) Mg, diethyl ether   (2) CO_2 (3) dil. HCl

(g) dil. HCl, heat

12.29 (a) [structure: cyclohexane ring with $CO_2^- Na^+$ groups]

(b) $C_6H_5CH_2O_2CCH_2CH_3$

(c) $CH_3CO_2^- + CH_3OH$ (d) $CH_3CO_2^-\ CH_3NH_3^+$ (e) $CH_3CO_2H + ClCH_2CO_2^-$

(f) $HO_2CCH_2CO_2^-\ Na^+$ (g) no reaction   (h) $C_6H_5CO_2CH_3$

12.30   (a) $C_6H_5CO_2^- \ Li^+ + H_2O$ (b) $C_6H_5O^- \ Li^+ + H_2O$

(c) no reaction. Hydroxides are not strong enough bases to remove the proton from an alcohol to any appreciable extent.

12.31 $CH_3(CH_2)_3CO_2H \rightleftharpoons CH_3(CH_2)_3CO_2^- + \quad H^+$
   0.200 – 0.00184 $\underline{M}$ 0.00184 $\underline{M}$ 0.00184 $\underline{M}$

$$\underline{K}_a = \frac{(0.00184)(0.00184)}{0.200 - 0.00184}$$
$$= 1.71 \times 10^{-5}$$

12.32   Mixture $\xrightarrow{NaHCO_3}$ $\underset{A}{C_6H_5CO_2^-} \ Na^+ \xrightarrow{NaOH}$

CH_3CH_2—⬡—$O^- \ Na^+ \xrightarrow{H_2O}$ no dissolved salts
                B   C

Solution D contains C_6H_5CHO.

12.33   (a) Shake with aqueous $NaHCO_3$; the acid dissolves and the ester does not.
       (b) Shake with aqueous NaOH; phenol dissolves and the ester does not.
       (c) Shake with aqueous NaOH; phenol dissolves and the alcohol does not.

12.34   (a) 90.08 (same as the molecular weight)
       (b) 64.05 (one-third the molecular weight)
       (c) 66.04 (one-half the molecular weight)

12.35   equivalents NaOH = $\underline{NV}$
                     = (0.307 eq./liter) (0.011 liter)
                     = 3.38×10^{-3}

$$\text{neut. eq. of acid} = \frac{0.250 \text{ g}}{3.38 \times 10^{-3} \text{ equivalents}}$$
$$= 74.0$$

12.36   Calculate the equivalent weights: (a) 88   (b) 74   (c) 52   (d) 73   (e) 72. Allowing for experimental error, (b), (d), and (e) are possibilities.

12.37   (a) $p\underline{K}_a = -\log (1.3 \times 10^{-4}) = 4 - \log 1.3 = 3.89$

       (b) $p\underline{K}_a = -\log (3.65 \times 10^{-5}) = 5 - \log 3.65 = 4.44$

12.38 $K_a = \dfrac{[H^+][A^-]}{[HA]}$ $pH = -\log [H^+] = 2.5$ $[H^+] = 10^{-2.5} = 3.2 \times 10^{-3}$

$K_a = \dfrac{(10^{-2.5})(10^{-2.5})}{0.010 - (3.2 \times 10^{-3})}$

$= \dfrac{10^{-5}}{0.0068} = 1.47 \times 10^{-3}$

12.39   A more electronegative substituent stabilizes the anion relative to the acid to a greater extent; therefore, FCH_2CO_2H is the strongest acid of the three and $BrCH_2CO_2H$ is the weakest.

12.40   (d), (h), (e), (f), (c), (b), (a), (g).

Note the order: alcohol (weakest), water, phenol, carbonic acid (weaker than RCO_2H because HCO_3^- reacts with RCO_2H), carboxylic acids (in order of increasing inductive effect).

12.41   (a) p-bromobenzoic acid   (b) m-bromobenzoic acid
(c) 3,5-dibromobenzoic acid

12.42   p-methylphenol, phenol, p-nitrophenol

p-Nitrophenol is a stronger acid than phenol because of resonance-stabilization of the anion, which is aided by the nitro group. (See Section 12.7G for the resonance structures.) p-Methylphenol is a weaker acid than phenol because the methyl group is electron-releasing and destabilizes the anion relative to the conjugate acid.

12.43   (b) is the most acidic, and (c) is the least acidic.

12.44   In each case, determine which of the parent acids is the weaker acid (Section 12.7). The anion of that acid is the stronger base.

(a) $CH_3CH{=}CH^-$ (b) $CH_3CO_2^-$ (c) $ClCH_2CO_2^-$ (d) $(CH_3)_3CO^-$

(e) $ClCH_2CH_2CO_2^-$ (f) $-CO_2^-$ (g) $C_6H_5CO_2^-$ (h) $C_6H_5CO_2^-$

In (f), the electron-releasing effect of two alkyl groups weakens an acid; therefore, cyclohexanecarboxylic acid is a weaker acid than heptanoic acid. In (g), the phenolic carboxylate is stabilized by hydrogen bonding; its conjugate acid is therefore stronger.

12.45   o-Phthalic acid has a smaller p$\underline{K}_1$ (more acidic) because intramolecular hydrogen bonding helps stabilize the anion.

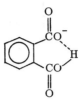

However, the p$\underline{K}_2$ of o-phthalic acid is larger because of (1) a stabilized reactant, and (2) a dianion that is destabilized by the proximity of two negative charges.

12.46   (a) $C_6H_5CO_2CH(CH_3)_2$ (b) $C_6H_5CO_2CH_2CH_2OH$ and $C_6H_5CO_2CH_2CH_2O_2CC_6H_5$

(c)

(d) no reaction   (e)

(f) (S)-$C_6H_5CO_2\overset{\displaystyle CH_3}{\underset{\displaystyle |}{C}}HCH_2CH_2CH_3$ (g) $C_6H_5\overset{O}{\overset{||}{C}}OCH_2CH_3$ and $C_6H_5\overset{^{18}O}{\overset{||}{C}}OCH_2CH_3$

In (g), the intermediate can lose either ^{16}O or ^{18}O:

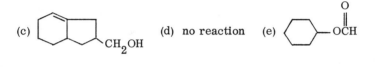

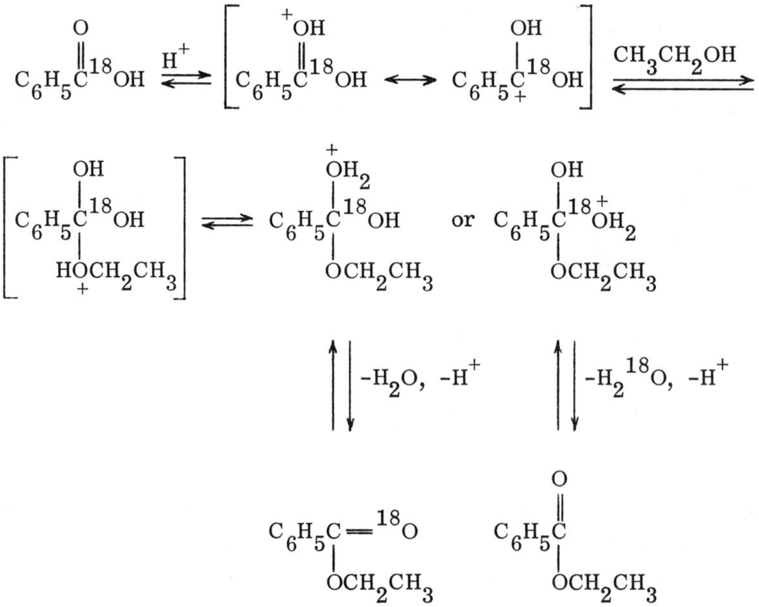

12.47   (b), (c), (a). The order is based upon decreasing steric hindrance.

12.48   (a) $CH_3CH_2CH_2OH$ (b) $CH_3CHClCH_2OH$ (c) $CH_2{=}CHCH_2OH$

(d) $HOCH_2CH_2CH_2OH$

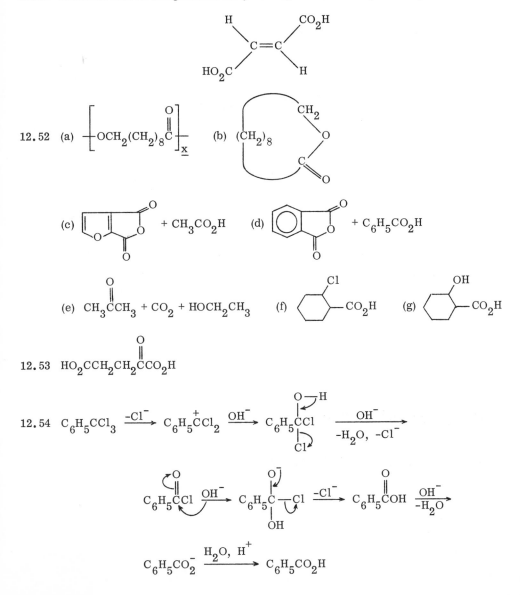

12.49 (a) $CH_3\overset{*}{C}HBrCO_2\overset{*}{C}HCH_2CH_3$ + $CH_3\overset{*}{C}HBrCO_2\overset{*}{C}HCH_2CH_3$

with CH₃ groups labeled above, and stereodescriptors:

(R)   (R)   (S)   (R)

(b) These two esters are diastereomers. The enantiomer of the first would be (S, S), while the enantiomer of the second would be (R, S).

12.50 (a) (structure)—$(CH_2)_4CH_3$ (b) $CH_3\overset{O}{\overset{\|}{C}}CH(CH_3)_2$ (c) (structure)=O

12.51 Fumaric acid is not geometrically arranged to form a cyclic anhydride.

12.52 (a) $\left[\!-OCH_2(CH_2)_8\overset{O}{\overset{\|}{C}}-\!\right]_x$ (b) $(CH_2)_8$ (cyclic structure with CH_2 and O)

(c) (structure) + CH_3CO_2H (d) (structure) + $C_6H_5CO_2H$

(e) $CH_3\overset{O}{\overset{\|}{C}}CH_3$ + CO_2 + $HOCH_2CH_3$ (f) (cyclohexane with Cl and CO_2H)   (g) (cyclohexane with OH and CO_2H)

12.53 $HO_2CCH_2CH_2\overset{O}{\overset{\|}{C}}CO_2H$

12.54 $C_6H_5CCl_3$ $\xrightarrow{-Cl^-}$ $C_6H_5\overset{+}{C}Cl_2$ $\xrightarrow{OH^-}$ $C_6H_5\overset{O-H}{\overset{|}{\underset{|}{\underset{Cl}{C}}}}Cl$ $\xrightarrow[-H_2O, \ -Cl^-]{OH^-}$

$C_6H_5\overset{O}{\overset{\|}{C}}Cl$ $\xrightarrow{OH^-}$ $C_6H_5\overset{O^-}{\overset{|}{\underset{|}{\underset{OH}{C}}}}Cl$ $\xrightarrow{-Cl^-}$ $C_6H_5\overset{O}{\overset{\|}{C}}OH$ $\xrightarrow[-H_2O]{OH^-}$

$C_6H_5CO_2^-$ $\xrightarrow{H_2O, \ H^+}$ $C_6H_5CO_2H$

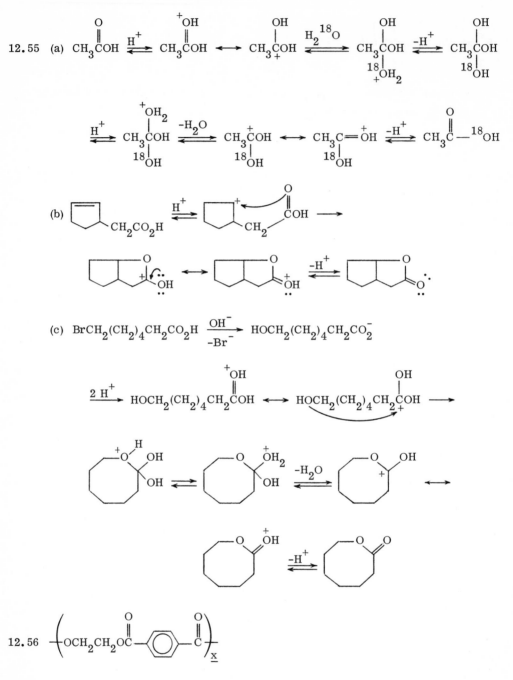

12.56 $\left(\text{OCH}_2\text{CH}_2\text{O}\overset{\text{O}}{\overset{\|}{\text{C}}}-\bigcirc-\overset{\text{O}}{\overset{\|}{\text{C}}}\right)_{\underline{x}}$

12.57 A: $CH_3CH(CO_2H)_2$ B: $CH_3CH_2CO_2H$

12.58 (a) $^{14}CH_3I \xrightarrow{CH_3CH_2CO_2^-}$ (b) $^{14}CH_3I \xrightarrow{KCN} {^{14}}CH_3CN \xrightarrow[\text{heat}]{H_2O,\ H^+}$

(c) $^{14}CH_3I \xrightarrow{OH^-}$ (d) $^{14}CH_3I \xrightarrow[AlX_3]{C_6H_6} C_6H_5{^{14}}CH_3 \xrightarrow[\text{heat}]{KMnO_4}$

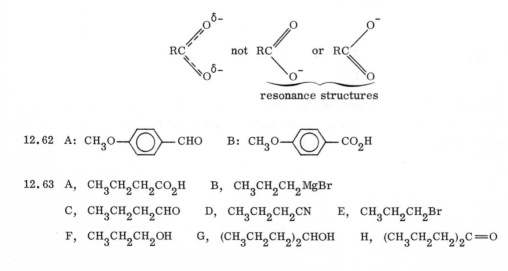

12.59 (a) $C_6H_5CO_2H \xrightarrow[FeBr_3]{Br_2}$ [structure: benzene ring with Br and $-CO_2H$] $\xrightarrow[(2)\ H_2O]{(1)\ LiAlH_4}$

(b) $C_6H_5CO_2H \xrightarrow[(2)\ H_2O]{(1)\ LiAlH_4} C_6H_5CH_2OH \xrightarrow{HCl} C_6H_5CH_2Cl \xrightarrow{CN^-}$

(c) $C_6H_5CO_2H \xrightarrow[H_2SO_4]{HNO_3}$ [structure: benzene ring with NO_2 and $-CO_2H$] $\xrightarrow[HCl]{Fe}$

[structure: benzene ring with $^+NH_3$ and $-CO_2H$] $\xrightarrow[(2)\ H_2O,\ OH^-]{(1)\ CH_3OH,\ heat}$ [structure: benzene ring with NH_2 and $-CO_2CH_3$]

12.60 A: $HO_2CCH_2CH_2CO_2H$ B: [structure: cyclic anhydride]   C: $CH_3O_2CCH_2CH_2CO_2CH_3$

D: $HOCH_2CH_2CH_2CH_2OH$

12.61 A carboxylate does not contain a true carbonyl group; it contains a composite of the two resonance structures.

[structure: RC with two O bonds showing $\delta-$, labeled] not [structure: RC with $=O$ and O^-] or [structure: RC with O^- and $=O$]

resonance structures

12.62 A: CH_3O-[benzene ring]$-CHO$ B: CH_3O-[benzene ring]$-CO_2H$

12.63 A, $CH_3CH_2CH_2CO_2H$ B, $CH_3CH_2CH_2MgBr$

C, $CH_3CH_2CH_2CHO$ D, $CH_3CH_2CH_2CN$ E, $CH_3CH_2CH_2Br$

F, $CH_3CH_2CH_2OH$ G, $(CH_3CH_2CH_2)_2CHOH$ H, $(CH_3CH_2CH_2)_2C=O$

Chapter 13

Derivatives of Carboxylic Acids

Some Important Features

The common derivatives of carboxylic acids are esters, amides, acid halides, acid anhydrides, and nitriles. These compounds all undergo addition-elimination reactions with reagents such as water, alcohols, ammonia, or amines. (See the chapter-end summary in the text.)

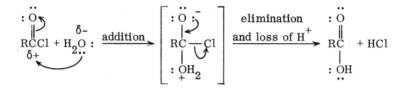

The reactivity of the carboxylic acid derivatives depends partly on the leaving group (Section 13.1).

$$\underset{\text{an acid chloride}}{\text{easier:}} \qquad \overset{O}{\overset{\|}{R\overset{}{C}Cl}} \quad + H_2O \longrightarrow R\overset{O}{\overset{\|}{C}}OH + HCl$$

$$\underset{\text{an ester}}{\text{more difficult:}} \qquad \overset{O}{\overset{\|}{R\overset{}{C}OR'}} + H_2O \underset{\text{heat}}{\overset{H^+}{\rightleftharpoons}} R\overset{O}{\overset{\|}{C}}OH + R'OH$$

In your study of the reactions of the carboxylic acid derivatives, pay particular attention to the similarities of the reactions and their mechanisms. The reactions of esters and amides, for example, are quite similar.

Simplified mechanisms:

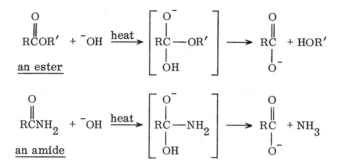

All carboxylic acid derivatives contain unsaturation and therefore can be reduced by catalytic hydrogenation or by LAH.

$$\underset{\text{an ester}}{\overset{\overset{\displaystyle O}{\|}}{RCOR'}} \quad \xrightarrow[\text{heat, pressure}]{H_2,\ Ni} \quad RCH_2OH + HOR'$$

Other topics covered in this chapter are α halogenation of acid halides, some other types of ester (lactones, polyesters, thioesters), some other types of amide (lactams, imides, carbamates, and others), and the reactions of nitriles.

Reminders

In writing mechanisms for the reactions of carboxylic acid derivatives, use electron dots for unshared electrons. They will help you see where protonation can occur and what the direction is of the electron shifts.

a protonated ester

Remember that protonation can occur in acidic solution, but not in alkaline solution.

You have probably noticed that we are using more-complex structures now than we did earlier in the text. Do not be intimidated by a complex structure. For example, under saponification conditions, the only reactive functional group in the following compound is an ester group.

a lactone, or ester, group

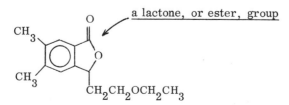

Answers to Problems

13.31 (a) p-bromobenzoyl bromide   (b) pentanoic propanoic anhydride

     (c) N,N-diethylbenzamide   (d) hexanenitrile

     (e) p-nitrophenyl acetate or p-nitrophenyl ethanoate

     (f) methyl p-nitrobenzoate

     (g) N-(p-bromophenyl)acetamide   or N-(p-bromophenyl)ethanamide

13.32 (a) [structure with $CH_2CH_2CH_3$]   (b) $CH_3CH_2\overset{O}{\overset{\|}{C}}N(CH_2CH_3)_2$ (c) $CH_3NH\overset{O}{\overset{\|}{C}}NHCH_3$

     (d) $CH_3CH_2CH_2\overset{CH_3}{\overset{|}{C}}HCN$ (e) $CH_3\overset{NH_2}{\overset{|}{C}}HCO_2CH_2CH_3$ (f) $CH_3CHClCH_2CN$

     (g) $C_6H_5\overset{O\ O}{\overset{\|\ \|}{C}}O\overset{}{C}H$

13.33 (b), (d), (a), (c), based primarily upon the difference in electronegativities between
the two atoms of the bond. The C—Cl bond has the greatest bond moment, partly
because of the polarizability of the Cl atom.

13.34 (a) [cyclohexyl]$-O_2CCH_3$ (b) $Br-$[phenyl]$-O_2CCH_3$ (c) $CH_3\overset{O}{\overset{\|}{C}}N$[ring]

     (d) $CH_3CO_2CH_2CH_3$ (e) $CH_3CO_2^-$ Na^+

In each reaction except (e), CH_3CO_2H (or $CH_3CO_2^-$) would also be an organic
product.

13.35 (a) [cyclohexyl]$-O_2CC_6H_5$ (b) $Br-$[phenyl]$-O_2CC_6H_5$ (c) $C_6H_5\overset{O}{\overset{\|}{C}}N$[ring]

     (d) $C_6H_5CO_2CH_2CH_3$ (e) $C_6H_5CO_2^-$ Na^+

13.36 (a) $C_6H_5CO_2^- + CH_3CH_2OH$ (b) $C_6H_5CO_2^- + H_2NCH_2CH_3$

     (c) $C_6H_5CO_2^-$ (d) $C_6H_5CO_2^- + CH_3CH_2CO_2^-$

     relative rates: (c) > (d) > (a) > (b)

13.37 (a) CH_3-[phenyl]$-CO_2CH_2CH_3 + CH_3OH$ (b) $Cl-$[phenyl]$-CO_2CH_2CH_3 + CH_3OH$

(c) $C_6H_5CO_2CH_2CH_3 + CH_3OH$

relative rates: (b) > (c) > (a)

Electron withdrawal by Cl in (b) decreases the electron density at the carbonyl carbon, thus increases the ease of nucleophilic attack by the ethoxide ion. Electron release by the methyl group in (a) has the opposite effect.

13.38 (a) $CH_3CO_2CH_2CH_3 \xrightarrow[\text{(2) } H^+]{\text{(1) } H_2O,\ OH^-,\ \text{heat}} CH_3CO_2H + CH_3CH_2OH$

(b) $CH_3CO_2CH_2CH_3 \xrightarrow[\text{(2) } H_2O]{\text{(1) } LiAlH_4} 2\ CH_3CH_2OH$

(c) $CH_3CO_2CH_2CH_3 \xrightarrow[\text{(2) } H_2O,\ H^+]{\text{(1) } 2\ CH_3MgI} (CH_3)_3COH$

(d) $CH_3CO_2CH_2CH_3 \xrightarrow[\text{(2) } SOCl_2]{\text{(1) } H_2O,\ H^+,\ \text{heat}} CH_3\overset{O}{\overset{\|}{C}}Cl \xrightarrow[AlCl_3]{C_6H_6} CH_3\overset{O}{\overset{\|}{C}}C_6H_5$

(e) $CH_3CO_2CH_2CH_3 \xrightarrow[\text{heat}]{NaOH,\ H_2O} CH_3CO_2^-\ Na^+ + HOCH_2CH_3$

(f) $CH_3CO_2CH_2CH_3 \xrightarrow{CH_3NH_2} CH_3\overset{O}{\overset{\|}{C}}NHCH_3 + HOCH_2CH_3$

13.39 Saponification is the reaction of choice because heating in dilute HCl would lead to addition of HCl and H_2O to the carbon–carbon double bond and allylic rearrangement.

13.40 (a) $(C_6H_5CH_2)_2Cd$ or $(C_6H_5CH_2)_2CuLi + Cl\overset{O}{\overset{\|}{C}}CH_2CH_3$

$C_6H_5\overset{O}{\overset{\|}{C}}Cl + Cd(CH_2CH_3)_2$ or $LiCu(CH_2CH_3)_2$

(b) $(CH_3)_2CH\overset{O}{\overset{\|}{C}}Cl + Cd(CH_2CH_3)_2$ or $LiCu(CH_2CH_3)_2$

$[(CH_3)_2CH]_2CuLi + Cl\overset{O}{\overset{\|}{C}}CH_2CH_3$

(c) $[(CH_3)_2CHCH_2C(CH_3)_2]_2CuLi + Cl\overset{\overset{\displaystyle O}{\|}}{C}C_6H_5$

$(CH_3)_2CHCH_2C(CH_3)_2\overset{\overset{\displaystyle O}{\|}}{C}Cl + Cd(C_6H_5)_2$ or $LiCu(C_6H_5)_2$

In (b), $[(CH_3)_2CH]_2Cd$ is unsuitable because 2° cadmium reagents are unstable. In the first set of reactants in (c), a similar restriction exists.

13.41 (a) $HO\overset{\overset{\displaystyle O}{\|}}{C}OH \longrightarrow H_2O + CO_2$ (b) $CH_3CH_2O\overset{\overset{\displaystyle O}{\|}}{C}OCH_2CH_3$

(c) $Cl\overset{\overset{\displaystyle O}{\|}}{C}OCH_2CH_3$ (d) $H_2N\overset{\overset{\displaystyle O}{\|}}{C}OCH_2CH_3$

Note that the more reactive group is the first to leave.

13.42 (a) $(\underline{R})-CH_3\overset{\overset{\displaystyle O}{\|}}{C}NH\overset{\overset{\displaystyle CH_3}{|}}{C}H(CH_2)_5CH_3$ (b) $C_6H_5\overset{\overset{\displaystyle O}{\|}}{C}-{}^{18}OCH(CH_3)_2$

13.43 The methyl ester would be saponified at a faster rate because of greater solubility in H_2O, and less hindrance in the transition state leading to the first intermediate (the rate-determining step).

13.44 The effect of the electron-releasing methyl groups on the benzene ring makes the carbonyl carbon atom less positive and less susceptible to attack.

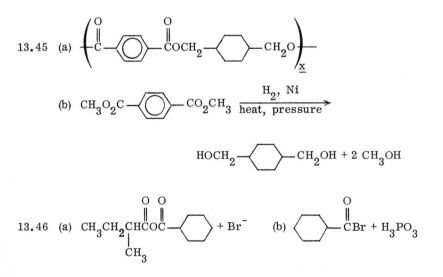

13.45 (a) $\left(\!\overset{\overset{\displaystyle O}{\|}}{C}\!-\!\bigcirc\!-\!\overset{\overset{\displaystyle O}{\|}}{C}OCH_2\!-\!\bigcirc\!-\!CH_2O\!\right)_{\!\!x}$

(b) $CH_3O_2C\!-\!\bigcirc\!-\!CO_2CH_3 \xrightarrow[\text{heat, pressure}]{H_2,\ Ni}$

$HOCH_2\!-\!\bigcirc\!-\!CH_2OH + 2\ CH_3OH$

13.46 (a) $CH_3CH_2\overset{\overset{\displaystyle CH_3}{|}}{C}H\overset{\overset{\displaystyle O}{\|}}{C}\overset{\overset{\displaystyle O}{\|}}{C}\!-\!\bigcirc + Br^-$ (b) $\bigcirc\!-\!\overset{\overset{\displaystyle O}{\|}}{C}Br + H_3PO_3$

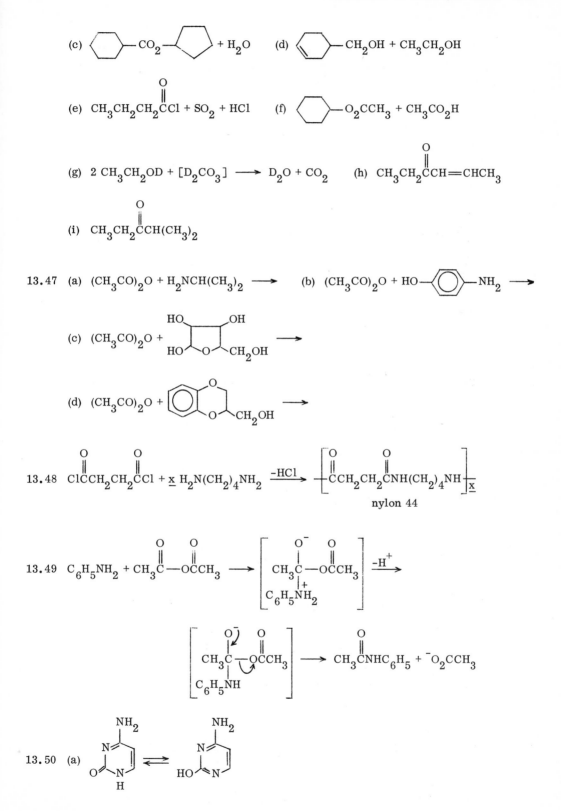

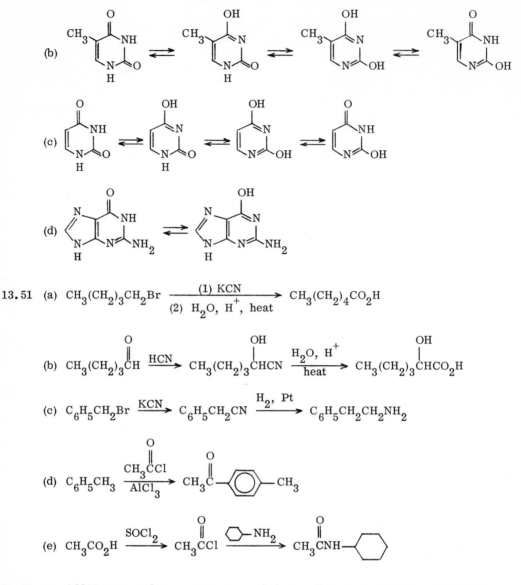

13.51 (a) $CH_3(CH_2)_3CH_2Br$ $\xrightarrow[\text{(2) } H_2O, H^+, \text{ heat}]{\text{(1) KCN}}$ $CH_3(CH_2)_4CO_2H$

(b) $CH_3(CH_2)_3\overset{O}{\overset{\|}{C}}H$ $\xrightarrow{\text{HCN}}$ $CH_3(CH_2)_3\overset{OH}{\overset{|}{C}}HCN$ $\xrightarrow[\text{heat}]{H_2O, H^+}$ $CH_3(CH_2)_3\overset{OH}{\overset{|}{C}}HCO_2H$

(c) $C_6H_5CH_2Br$ $\xrightarrow{\text{KCN}}$ $C_6H_5CH_2CN$ $\xrightarrow{H_2, \text{ Pt}}$ $C_6H_5CH_2CH_2NH_2$

(d) $C_6H_5CH_3$ $\xrightarrow[\text{AlCl}_3]{CH_3\overset{O}{\overset{\|}{C}}Cl}$ $CH_3\overset{O}{\overset{\|}{C}}$—⟨ ⟩—$CH_3$

(e) CH_3CO_2H $\xrightarrow{SOCl_2}$ $CH_3\overset{O}{\overset{\|}{C}}Cl$ $\xrightarrow{⟨ ⟩-NH_2}$ $CH_3\overset{O}{\overset{\|}{C}}NH$—⟨ ⟩

13.52 (a) Add $NaHCO_3$ solution. Benzoic acid shows effervescence (CO_2); methyl ben-
       zoate does not.
   (b) Subject the samples to alkaline hydrolysis; the amide will release fumes of
       the basic $CH_3CH_2NH_2$, which turn moist litmus paper blue.
   (c) Add $AgNO_3$ solution; C_6H_5COCl gives a AgCl precipitate.

13.53 (a) $CH_3CH_2CH_2OH + HO\overset{CH_3}{\overset{|}{C}}HCH_2CH_2CH_3$ (b) ⟨ ⟩—CO_2H

   (c) $CH_3CH_2CH_2NHCH_3$ (d) ⟨ ⟩—$CH_2OH + CH_3OH$

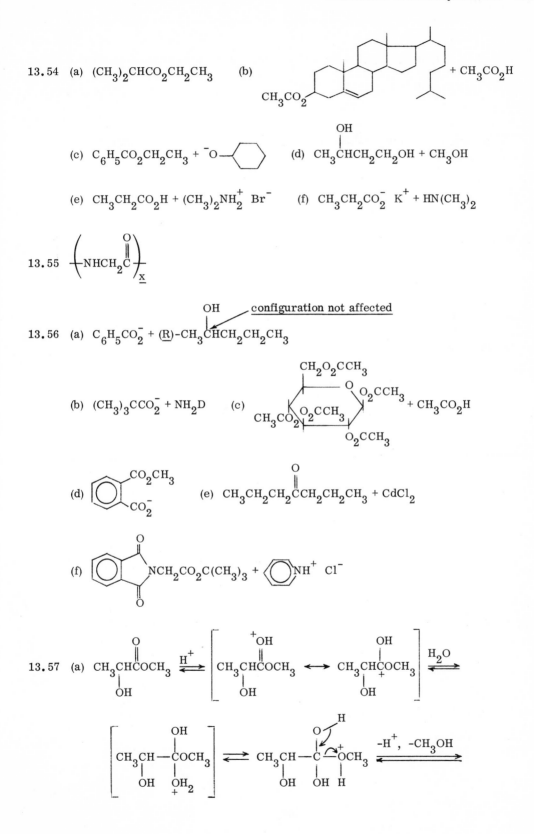

13.54 (a) $(CH_3)_2CHCO_2CH_2CH_3$ (b) $+ CH_3CO_2H$

(c) $C_6H_5CO_2CH_2CH_3 + {}^-O-$⬡   (d) $CH_3\overset{OH}{\underset{|}{C}}HCH_2CH_2OH + CH_3OH$

(e) $CH_3CH_2CO_2H + (CH_3)_2\overset{+}{N}H_2 \ Br^-$ (f) $CH_3CH_2CO_2^- \ K^+ + HN(CH_3)_2$

13.55 $\left(-NHCH_2\overset{\overset{O}{\|}}{C}-\right)_{\underline{x}}$

13.56 (a) $C_6H_5CO_2^- + (\underline{R})-CH_3\overset{OH}{\underset{|}{C}}HCH_2CH_2CH_3$ ← configuration not affected

(b) $(CH_3)_3CCO_2^- + NH_2D$ (c) $+ CH_3CO_2H$

(d) ⬡$\overset{CO_2CH_3}{\underset{CO_2^-}{}}$ (e) $CH_3CH_2CH_2\overset{\overset{O}{\|}}{C}CH_2CH_2CH_3 + CdCl_2$

(f) [phthalimide]$NCH_2CO_2C(CH_3)_3 + $⬡$\overset{+}{N}H \ Cl^-$

13.57 (a) $CH_3\overset{\overset{O}{\|}}{C}H\overset{}{C}OCH_3$ with $\overset{OH}{\underset{|}{}}$ $\xrightarrow{H^+}$ $\left[CH_3\overset{\overset{+OH}{\|}}{C}H\overset{}{C}OCH_3 \longleftrightarrow CH_3\overset{OH}{\underset{|}{C}}H\overset{}{\underset{+}{C}}OCH_3\right]$ with $\overset{OH}{\underset{|}{}}$ $\overset{H_2O}{\underset{}{\rightleftarrows}}$

$\left[CH_3CH-\overset{OH}{\underset{\overset{OH_2}{+}}{C}}OCH_3\right]$ $\rightleftarrows$ $CH_3CH-\overset{\overset{O}{\diagup}H}{\underset{\overset{|}{OH} \ \overset{|}{OH} \ H}{C}}\overset{+}{O}CH_3$ $\xrightarrow{-H^+, \ -CH_3OH}$

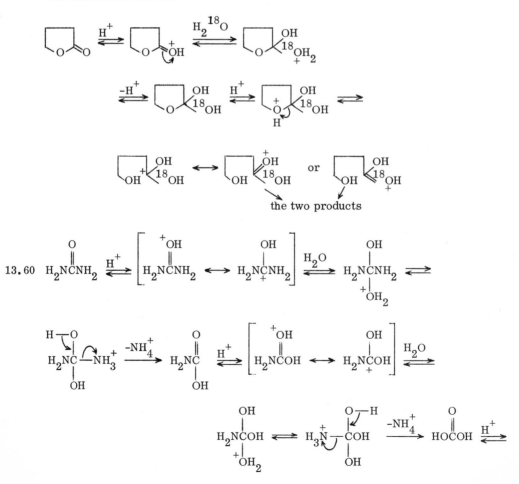

$$CH_3CHCOH \quad \text{plus some} \quad CH_2{=}CHCO_2H \quad \text{by a dehydration reaction}$$

(with O double bond on the carbonyl carbon and OH on the CH)

(b) $(CH_3)_2CHCOCH_3 \xrightarrow{OH^-} \left[(CH_3)_2CHC \overset{O^-}{\underset{O}{\underset{H}{\big|}}} OCH_3 \right] \xrightarrow{-CH_3OH} (CH_3)_2CHCO^-$

13.58   (ethylene carbonate structure) $=O$ and $\left(\!\!\begin{array}{c} O \\ \| \\ COCH_2CH_2O \end{array}\!\!\right)_{\underline{x}}$

13.59 $HOCH_2CH_2CH_2\overset{^{18}O}{\overset{\|}{C}}OH + HOCH_2CH_2CH_2\overset{O}{\overset{\|}{C}}{}^{18}OH$

The mechanism follows:

(series of cyclic lactone mechanism structures with ^{18}O labeling)

13.60 $H_2N\overset{O}{\overset{\|}{C}}NH_2 \xrightarrow{H^+} \left[H_2N\overset{^+OH}{\overset{\|}{C}}NH_2 \longleftrightarrow H_2N\overset{OH}{\underset{+}{\overset{|}{C}}NH_2} \right] \xrightarrow{H_2O} H_2N\overset{OH}{\underset{\overset{|}{+}OH_2}{\overset{|}{C}}NH_2} \rightleftharpoons$

$H_2N\overset{H\frown O}{\underset{OH}{\overset{|}{C}}}NH_3^+ \xrightarrow{-NH_4^+} H_2N\overset{O}{\underset{OH}{\overset{\|}{C}}} \xrightarrow{H^+} \left[H_2N\overset{^+OH}{\overset{\|}{C}}OH \longleftrightarrow H_2N\overset{OH}{\underset{+}{\overset{|}{C}}OH} \right] \xrightarrow{H_2O}$

$H_2N\overset{OH}{\underset{\overset{|}{+}OH_2}{\overset{|}{C}}OH} \rightleftharpoons H_3\overset{+}{N}\overset{O\frown H}{\underset{OH}{\overset{|}{C}}OH} \xrightarrow{-NH_4^+} HO\overset{O}{\overset{\|}{C}}OH \xrightarrow{H^+}$

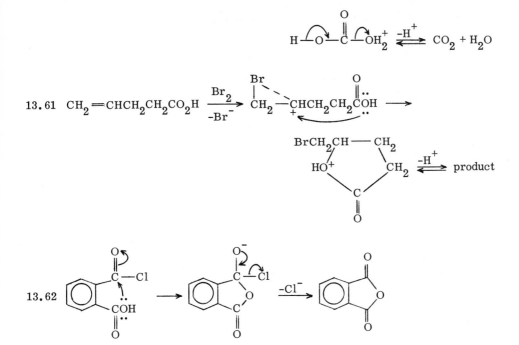

$$H-O-\overset{\overset{\displaystyle O}{\|}}{C}-\overset{+}{O}H_2 \xrightarrow{-H^+} CO_2 + H_2O$$

13.61 $CH_2=CHCH_2CH_2CO_2H \xrightarrow[-Br^-]{Br_2}$ product

13.62

13.63   When A is hydrolyzed, the ^{18}O—R bond is not broken.

$$CH_3\overset{\overset{\displaystyle O}{\|}}{C}-^{18}OCH_2CH_3 \underset{}{\overset{H_2O,\ H^+}{\rightleftharpoons}} \left[CH_3\overset{\overset{\displaystyle OH}{|}}{\underset{\underset{\displaystyle OH}{|}}{C}}-^{18}OCH_2CH_3 \right] \rightleftharpoons$$

$$CH_3\overset{\overset{\displaystyle O}{\|}}{C}OH + H^{18}OCH_2CH_3$$

When B is subjected to hydrolysis, however, the t-butyl group can break off as a carbocation, leading to a mixture of products. One possibility for a carbocation reaction follows:

$$CH_3\overset{\overset{\displaystyle O}{\|}}{C}-^{18}O-C(CH_3)_3 \rightleftharpoons \left[CH_3\overset{\overset{\displaystyle O}{\|}}{C}-^{18}O^- \longleftrightarrow CH_3\overset{\overset{\displaystyle O^-}{|}}{C}=^{18}O \right] + \left[\overset{+}{C}(CH_3)_3 \right]$$

$$\rightleftharpoons CH_3\overset{\overset{\displaystyle OC(CH_3)_3}{|}}{C}=^{18}O \underset{}{\overset{H_2O,\ H^+}{\rightleftharpoons}} CH_3\overset{\overset{\displaystyle ^{18}O}{\|}}{C}OH + HOC(CH_3)_3$$

13.64 $H\overset{\overset{\displaystyle O}{\|}}{C}CH_2CH_2CO_2H \xrightarrow{HCN} \overset{\overset{\displaystyle OH}{|}}{\underset{\underset{\displaystyle CN}{|}}{C}}HCH_2CH_2CO_2H \underset{}{\overset{-H_2O}{\rightleftharpoons}} \xrightarrow{-H_2O}$

13.65 $CH_3CH_2CH_2Br \xrightarrow[(1)]{NaCN} CH_3CH_2CH_2CN \xrightarrow[(2)]{hydrolysis}$

$\underline{A}$

$CH_3CH_2CH_2CO_2H \xrightarrow[(3)]{SOCl_2} CH_3CH_2CH_2\overset{\displaystyle O}{\overset{\displaystyle \|}{C}}Cl$

$\underline{B}$ $\underline{C}$

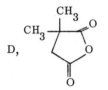

$\xrightarrow[(4)]{NaOH} CH_3CH_2CH_2CO_2^- \xrightarrow{CH_3CH_2CH_2\overset{\displaystyle O}{\overset{\displaystyle \|}{C}}Cl} \left(CH_3CH_2CH_2\overset{\displaystyle O}{\overset{\displaystyle \|}{C}} \right)_2 O$

$\underline{D}$

13.66   A, $(CH_3)_2C{=}CHCO_2H$ B, $(CH_3)_2\overset{\displaystyle CN}{\overset{\displaystyle |}{C}}CH_2CO_2H$ C, $(CH_3)_2\overset{\displaystyle CO_2H}{\overset{\displaystyle |}{C}}CH_2CO_2H$

D,

13.67   A, $C_6H_5CH_2CO_2H$ B, $C_6H_5CH_2CO_2CH_3$ C, $C_6H_5CH_2CH_2OH$

D, $C_6H_5CH_2\overset{\displaystyle O}{\overset{\displaystyle \|}{C}}Cl$ E, $C_6H_5CH_2\overset{\displaystyle O}{\overset{\displaystyle \|}{C}}NH_2$

Chapter 14

Enolates and Carbanions:
Building Blocks for Organic Synthesis

Some Important Features

    A hydrogen atom α to a carbonyl group is slightly acidic and may be removed by a base. The reason for this acidity of carbonyl compounds is the resonance-stabilization of the enolate ion. The extent of enolate formation is determined by the acidity of the carbonyl compound and by the strength of the base.

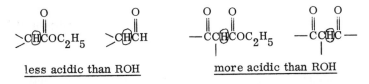

less acidic than ROH   more acidic than ROH

$$CH_3\overset{O}{\overset{\|}{C}}CH_2\overset{O}{\overset{\|}{C}}CH_3 + {}^-OC_2H_5 \rightleftharpoons CH_3\overset{O}{\overset{\|}{C}}\bar{C}H\overset{O}{\overset{\|}{C}}CH_3 + HOC_2H_5$$

    Enolate ions are useful as synthetic intermediates. For example, enolate ions can act as nucleophiles in S_N2 reactions with alkyl halides.

An alkylation reaction:

$$CH_3\overset{O}{\overset{\|}{C}}\bar{C}H\overset{O}{\overset{\|}{C}}CH_3 + R-X \xrightarrow{S_N2} CH_3\overset{O}{\overset{\|}{C}}\underset{\underset{R}{|}}{CH}\overset{O}{\overset{\|}{C}}CH_3 + X^-$$

Enolate ions can also attack carbonyl groups.

An aldol condensation:

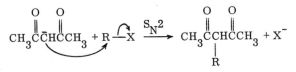

a β-hydroxy aldehyde

An ester condensation:

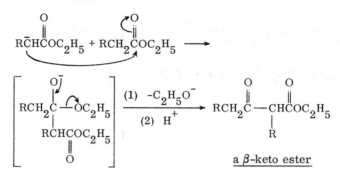

Enolate ions can also attack an α,β-unsaturated carbonyl compound in a 1,4-addition reaction (Michael addition).

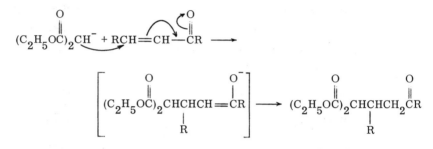

The products of enolate reactions may be subjected to further reactions such as a second enolate reaction, hydrolysis, decarboxylation, or dehydration. These topics are discussed in the text.

Reminders

In predicting the products of an enolate or related reaction, look first for the most acidic hydrogen. Then look for the most likely target of nucleophilic attack – for example, a carbon-halogen bond or a carbonyl group.

most acidic   nucleophilic

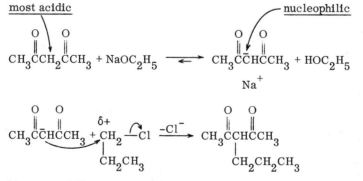

attacking a partially positive carbon

In solving problems asking for a synthetic route, first determine the type of product. Is the product a substituted acetic acid? (Try a malonic ester alkylation.) A substituted acetone? (Try an acetoacetic ester alkylation.) An α,β-unsaturated aldehyde? (Try an aldol condensation.) A β-keto ester? (Try an ester condensation.) After you deduce a likely route, use your pencil to divide the structure into its pieces.

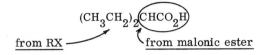

from RX ──── from malonic ester

At this point, you can work the problem backwards.

$(CH_3CH_2)_2C{\displaystyle \atop \displaystyle}\Big\langle {}^{CO_2H}_{CO_2H}$ $\xrightarrow{heat}$ product

↑ | H_2O, H^+, heat

$(CH_3CH_2)_2C{\displaystyle \atop \displaystyle}\Big\langle {}^{CO_2C_2H_5}_{CO_2C_2H_5}$

↑ | (1) $NaOC_2H_5$
  | (2) CH_3CH_2Br

$CH_3CH_2CH(CO_2C_2H_5)_2$ $\xleftarrow{CH_3CH_2Br}$ $^-CH(CO_2C_2H_5)_2$ $\xleftarrow{NaOC_2H_5}$ $CH_2(CO_2C_2H_5)_2$

Answers to Problems

14.28 (a) $C_6H_5\overset{\underset{\textstyle CH_3}{|}}{C}HCO_2C_2H_5$ (b) $CH_3\overset{\underset{}{\overset{O}{\|}}}{C}CH_2CN$ (c) $OHCCH_2CH{=}CHCHO$ with CH_3 on circled position

(d) $(CH_3)_3CCH_2CO_2H$ (e) $CH_3CH_2NO_2$ (f) $CH_3CCl_2CHCl_2$

In (f), the circled proton is acidic because of the cumulative inductive effect of the chlorine atoms.

14.29 (a) $C_2H_5O_2CCH_2CN + C_2H_5O^- \rightleftharpoons C_2H_5O_2C\bar{C}HCN + C_2H_5OH$

(b) $C_6H_5CH_2CN + C_2H_5O^- \rightleftharpoons C_6H_5\bar{C}HCN + C_2H_5OH$

(c) [cyclohexane-1,3-dione] $+ C_2H_5O^- \rightleftharpoons$ [cyclohexane-1,3-dione anion] $- + C_2H_5OH$

(d) $CH_3CH_2\overset{\underset{}{\overset{O}{\|}}}{C}\overset{\underset{\textstyle CH_3}{|}}{C}H\overset{\underset{}{\overset{O}{\|}}}{C}CH_2CH_3 + C_2H_5O^- \rightleftharpoons CH_3CH_2\overset{\underset{}{\overset{O}{\|}}}{C}\overset{\underset{\textstyle CH_3}{|}}{\bar{C}}\overset{\underset{}{\overset{O}{\|}}}{C}CH_2CH_3 + C_2H_5OH$

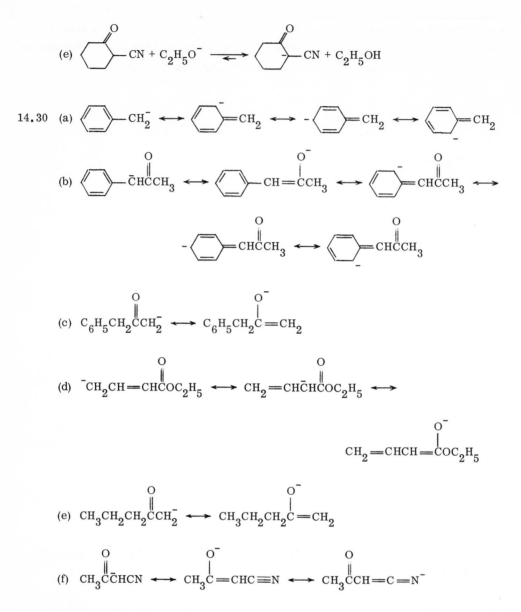

14.31   (b), (a), (e), (d), (c), (f)

The compounds with hydrogens α to only one carbonyl group, (b) and (a), are less acidic than ethanol (e), while those with hydrogens α to two carbonyl groups are more acidic. None of these other compounds is as acidic as a carboxylic acid (f). Also note that a hydrogen α to an aldehyde group is slightly more acidic than one α to an ester group.

14.32   (a) $\rightleftharpoons$ $CH_3CO_2C_2H_5 + CH_3\overset{\overset{\displaystyle O}{\|}}{C}\overset{-}{C}HCO_2C_2H_5$

(b) $\rightleftharpoons$ $CH_3\overset{O}{\overset{\|}{C}}CHNO_2 + H_2O$

(c) $\rightleftharpoons$ 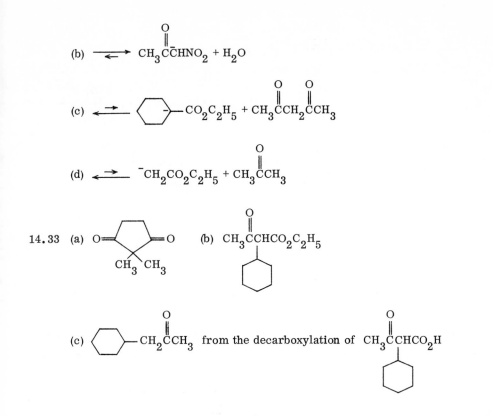 $-CO_2C_2H_5 + CH_3\overset{O}{\overset{\|}{C}}CH_2\overset{O}{\overset{\|}{C}}CH_3$

(d) $\rightleftharpoons$ $^-CH_2CO_2C_2H_5 + CH_3\overset{O}{\overset{\|}{C}}CH_3$

14.33 (a) $O=$⬠$=O$ (b) $CH_3\overset{O}{\overset{\|}{C}}CHCO_2C_2H_5$
       CH_3 CH_3

(c) ⬡ $-CH_2\overset{O}{\overset{\|}{C}}CH_3$ from the decarboxylation of $CH_3\overset{O}{\overset{\|}{C}}CHCO_2H$

(d) The attacking nucleophile is $^-CH(CO_2C_2H_5)$, and the leaving group is the tosylate ion (^-OTs). The reaction proceeds by an S_N2 mechanism (backside displacement), resulting in the <u>trans</u> product:

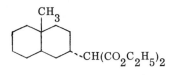

$-CH(CO_2C_2H_5)_2$

(e) $CH_2[CH(CO_2C_2H_5)_2]_2$

(f) $HO_2CCH_2CH_2CH_2CO_2H$ from the decarboxylation of $(HO_2C)_2CHCH_2CH(CO_2H)_2$

(g) $CH_3\overset{O}{\overset{\|}{C}}CHCO_2C_2H_5$ (h) $CH_3\overset{O}{\overset{\|}{C}}CH_2CH_2\overset{O}{\overset{\|}{C}}CH_3$
       $CH_2\overset{}{C}CH_3$
       $\overset{\|}{O}$

14.34 (a) $CH_2(CO_2C_2H_5)_2$ $\xrightarrow[\text{(2) }⬠-CH_2Br]{\text{(1) NaOC}_2H_5}$ ⬠ $-CH_2CH(CO_2C_2H_5)_2$ $\xrightarrow[\text{heat}]{H_2O,\ H^+}$

(b) $CH_2(CO_2C_2H_5)_2$ $\xrightarrow[\text{(2) } CH_3CHBrCO_2C_2H_5]{\text{(1) } NaOC_2H_5}$

$$C_2H_5O_2CCHCH(CO_2C_2H_5)_2 \xrightarrow[\text{heat}]{H_2O, \ H^+}$$

with CH_3 group on the CH.

(c) $CH_3\overset{O}{\overset{\|}{C}}CH_2CO_2C_2H_5$ $\xrightarrow[\text{(2) } C_6H_5CH_2Br]{\text{(1) } NaOC_2H_5}$ $CH_3\overset{O}{\overset{\|}{C}}CHCO_2C_2H_5$ $\xrightarrow[\text{heat}]{H_2O, \ H^+}$
with $CH_2C_6H_5$ group on the CH.

(d) $CH_2(CO_2C_2H_5)_2$ $\xrightarrow[\text{(2) } O=\text{(cyclopentene)}]{\text{(1) } NaOC_2H_5}$ $O=\text{(cyclopentane ring)}-\bar{C}(CO_2C_2H_5)_2$ $\xrightarrow[\text{heat}]{H_2O, \ H^+}$

14.35   solvent, CH_3OH; base, $NaOCH_3$; RX, $(CH_3)_2CHBr$

14.36   <u>n</u>-Pentyl bromide can undergo elimination when treated with a bulky base. Therefore, this alkyl halide should be used for the first substitution with the less bulky
$^-CH(CO_2C_2H_5)_2$, rather than with $CH_3\bar{C}(CO_2C_2H_5)_2$.
    A second reason for using <u>n</u>-pentyl bromide first is that disubstitution is
less likely to occur with <u>n</u>-pentyl bromide than with CH_3I.

14.37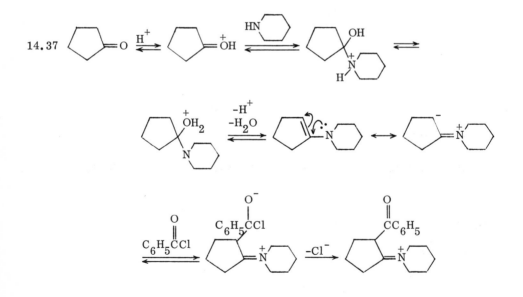

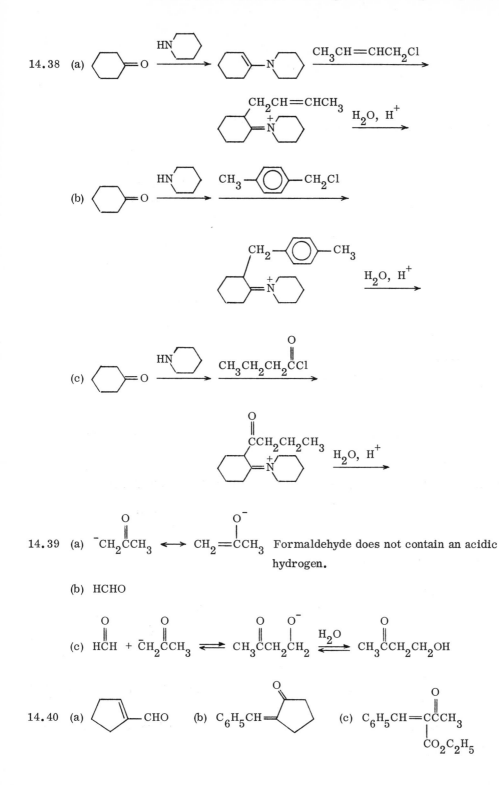

14.38   (a)

(b)

(c)

14.39   (a) $^{-}CH_2\overset{O}{\overset{\|}{C}}CH_3$ $\longleftrightarrow$ $CH_2=\overset{O^-}{\overset{|}{C}}CH_3$ Formaldehyde does not contain an acidic hydrogen.

(b)   HCHO

(c) $H\overset{O}{\overset{\|}{C}}H$ + $^{-}CH_2\overset{O}{\overset{\|}{C}}CH_3$ $\rightleftharpoons$ $CH_3\overset{O}{\overset{\|}{C}}CH_2\overset{O^-}{\overset{|}{C}}H_2$ $\overset{H_2O}{\rightleftharpoons}$ $CH_3\overset{O}{\overset{\|}{C}}CH_2CH_2OH$

14.40   (a)   [cyclopentene]—CHO   (b) $C_6H_5CH=$ [cyclopentanone]   (c) $C_6H_5CH=\overset{O}{\overset{\|}{C}}\underset{\overset{|}{CO_2C_2H_5}}{C}CH_3$

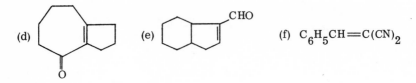

(f) $C_6H_5CH=C(CN)_2$

The mechanism for (d) follows:

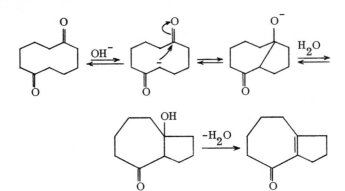

14.41 (a) $=O + C_6H_5CH_2CO_2C_2H_5 \xrightarrow{\text{NaOC}_2\text{H}_5}$

(b) $C_6H_5CHO + CH_3\overset{\overset{\displaystyle O}{\|}}{C}C_6H_5 \underset{\longleftarrow}{\overset{OH^-}{\longrightarrow}}$ (c) $-CHO + CH_3NO_2 \xrightarrow{OH^-}$

(d) $C_6H_5CHO + C_6H_5CH_2CO_2C_2H_5 \xrightarrow{\text{NaOC}_2\text{H}_5} C_6H_5CH=\underset{\underset{\displaystyle C_6H_5}{|}}{C}CO_2C_2H_5$

(e) $C_6H_5CH_2CO_2H \xrightarrow[\text{heat}]{CH_3CH_2OH, \ H^+} C_6H_5CH_2CO_2C_2H_5$

$\xrightarrow[\text{(2) }(CH_3)_2C=O]{\text{(1) NaOC}_2\text{H}_5} (CH_3)_2C=\underset{\underset{\displaystyle C_6H_5}{|}}{C}CO_2C_2H_5$

14.42 (a) $+$ (b) $(CH_3)_3CCO_2H + (CH_3)_3CCH_2OH$

Neither starting aldehyde contains an α hydrogen; both undergo Cannizzaro reactions.

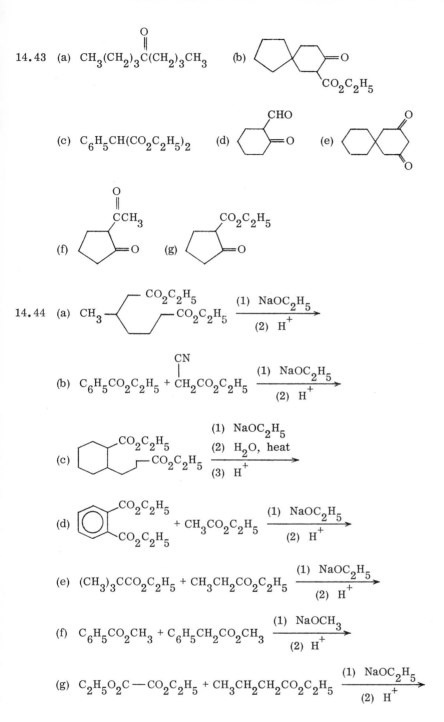

14.43 (a) $CH_3(CH_2)_3\overset{\overset{\displaystyle O}{\|}}{C}(CH_2)_3CH_3$ (b)

(c) $C_6H_5CH(CO_2C_2H_5)_2$ (d)   (e)

(f)   (g)

14.44 (a) CH_3 $-CO_2C_2H_5$ $-CO_2C_2H_5$

(1) $NaOC_2H_5$
(2) H^+

(b) $C_6H_5CO_2C_2H_5$ + $\overset{\overset{\displaystyle CN}{|}}{CH_2}CO_2C_2H_5$

(1) $NaOC_2H_5$
(2) H^+

(c)

(1) $NaOC_2H_5$
(2) H_2O, heat
(3) H^+

(d)

(1) $NaOC_2H_5$
(2) H^+

+ $CH_3CO_2C_2H_5$

(e) $(CH_3)_3CCO_2C_2H_5$ + $CH_3CH_2CO_2C_2H_5$

(1) $NaOC_2H_5$
(2) H^+

(f) $C_6H_5CO_2CH_3$ + $C_6H_5CH_2CO_2CH_3$

(1) $NaOCH_3$
(2) H^+

(g) $C_2H_5O_2C-CO_2C_2H_5$ + $CH_3CH_2CH_2CO_2C_2H_5$

(1) $NaOC_2H_5$
(2) H^+

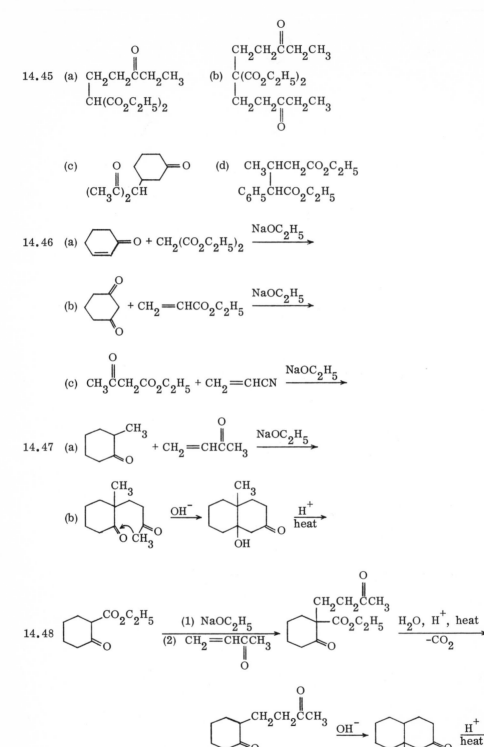

14.49 (a) $CH_3CO_2H + CH_3CH_2OH \xrightarrow[\longleftarrow]{\text{dil. HCl, heat}} CH_3CO_2C_2H_5 + H_2O$

(b) $C_2H_5OH \xrightarrow{\text{Na}} C_2H_5ONa \xrightarrow[\text{(2) } H^+]{\text{(1) } CH_3CO_2C_2H_5} CH_3\overset{\overset{\text{O}}{\|}}{C}CH_2CO_2C_2H_5$

(c) $CH_3CH_2OH \xrightarrow{PBr_3} CH_3CH_2Br \xrightarrow[\text{NaOC}_2H_5]{CH_3\overset{\overset{\text{O}}{\|}}{C}CH_2CO_2C_2H_5}$

$CH_3\overset{\overset{\text{O}}{\|}}{C}\underset{\underset{CH_2CH_3}{|}}{C}HCO_2C_2H_5 \xrightarrow[\text{heat}]{\text{dil. HCl}} CH_3\overset{\overset{\text{O}}{\|}}{C}CH_2CH_2CH_3$

(d) $CH_2(CO_2C_2H_5)_2 \xrightarrow[\text{(2) } 2\ CH_3CH_2Br]{\text{(1) } 2\ NaOC_2H_5} (CH_3CH_2)_2C(CO_2C_2H_5)_2$

$\xrightarrow[\text{heat}]{\text{dil. HCl}} (CH_3CH_2)_2CHCO_2H$

(e) $CH_3CO_2H \xrightarrow[Br_2]{PBr_3} BrCH_2\overset{\overset{\text{O}}{\|}}{C}Br \xrightarrow{CH_3CH_2OH} BrCH_2CO_2C_2H_5$

14.50 (R)-$C_6H_5\overset{\overset{OH}{|}}{C}HCO_2C_2H_5 \xrightarrow[\longleftarrow]{OH^-}$ achiral $\left[C_6H_5\overset{\overset{OH}{|}}{\underset{-}{C}}CO_2C_2H_5 \longleftrightarrow C_6H_5\overset{\overset{HO}{|}}{C}=\overset{\overset{O^-}{|}}{C}OC_2H_5 \right]$

$\xrightarrow[\longleftarrow]{H_2O,\ -OH^-}$ (R) and (S)-$C_6H_5\overset{\overset{OH}{|}}{C}HCO_2C_2H_5$

14.51 $CH_3CHO \xrightarrow{CO_3^{2-}} {}^-CH_2CHO \xrightarrow[\longleftarrow]{HCHO} {}^-OCH_2CH_2CHO \xrightarrow[\longleftarrow]{H_2O}$

$HOCH_2CH_2CHO \xrightarrow[\longleftarrow]{CO_3^{2-}} HOCH_2\bar{C}HCHO \xrightarrow{HCHO}$

$HOCH_2\overset{\overset{CH_2O^-}{|}}{C}HCHO \xrightarrow[\longleftarrow]{H_2O} HOCH_2\overset{\overset{CH_2OH}{|}}{C}HCHO \xrightarrow{CO_3^{2-}}$

$(HOCH_2)_2\bar{C}CHO \xrightarrow{HCHO} (HOCH_2)_2\overset{\overset{CH_2O^-}{|}}{C}CHO \xrightarrow[\longleftarrow]{H_2O}$ product

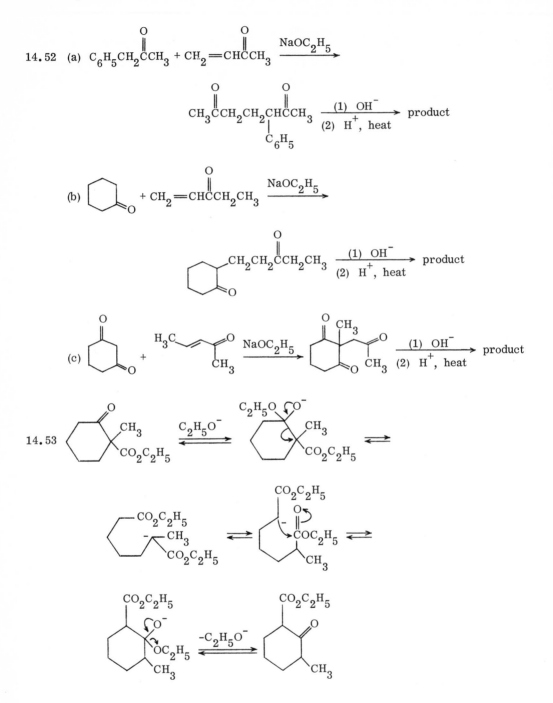

14.54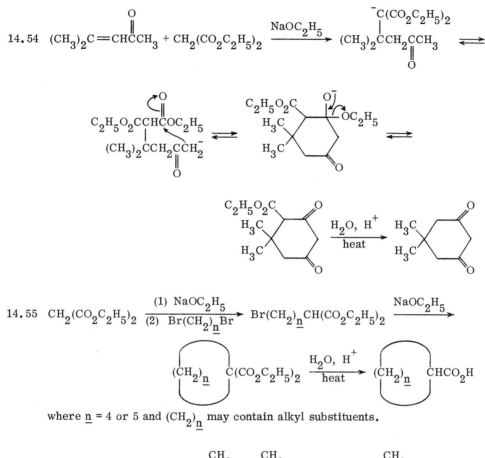

14.55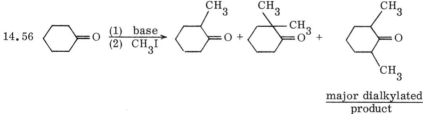

where $\underline{n}$ = 4 or 5 and $(CH_2)_{\underline{n}}$ may contain alkyl substituents.

14.56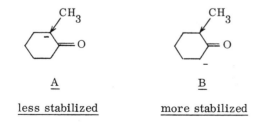

major dialkylated
product

Of the two anions (A and B) leading to dialkylated products, A is in the lowest concentration because of the electron-releasing effect of the methyl group.

A

less stabilized

B

more stabilized

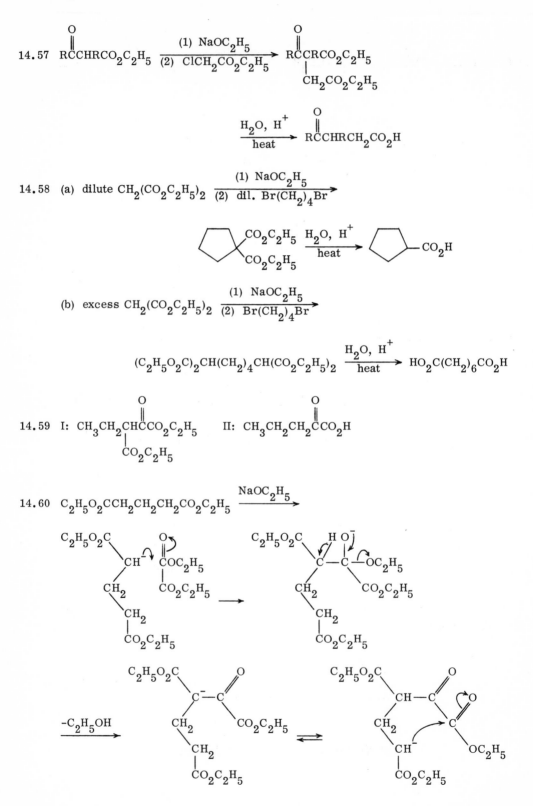

14.57 $\underset{\text{O}}{\overset{\text{O}}{\text{R}\text{C}\text{CHRCO}_2\text{C}_2\text{H}_5}}$ $\xrightarrow[\text{(2) ClCH}_2\text{CO}_2\text{C}_2\text{H}_5]{\text{(1) NaOC}_2\text{H}_5}$ $\underset{\text{CH}_2\text{CO}_2\text{C}_2\text{H}_5}{\overset{\text{O}}{\text{R}\text{C}\text{CRCO}_2\text{C}_2\text{H}_5}}$

$\xrightarrow[\text{heat}]{\text{H}_2\text{O, H}^+}$ $\overset{\text{O}}{\text{R}\text{C}\text{CHRCH}_2\text{CO}_2\text{H}}$

14.58 (a) dilute $\text{CH}_2(\text{CO}_2\text{C}_2\text{H}_5)_2$ $\xrightarrow[\text{(2) dil. Br(CH}_2)_4\text{Br}]{\text{(1) NaOC}_2\text{H}_5}$

$\xrightarrow[\text{heat}]{\text{H}_2\text{O, H}^+}$

(b) excess $\text{CH}_2(\text{CO}_2\text{C}_2\text{H}_5)_2$ $\xrightarrow[\text{(2) Br(CH}_2)_4\text{Br}]{\text{(1) NaOC}_2\text{H}_5}$

$(\text{C}_2\text{H}_5\text{O}_2\text{C})_2\text{CH(CH}_2)_4\text{CH(CO}_2\text{C}_2\text{H}_5)_2$ $\xrightarrow[\text{heat}]{\text{H}_2\text{O, H}^+}$ $\text{HO}_2\text{C(CH}_2)_6\text{CO}_2\text{H}$

14.59 I: $\underset{\text{CO}_2\text{C}_2\text{H}_5}{\overset{\text{O}}{\text{CH}_3\text{CH}_2\text{CH}\text{C}\text{CO}_2\text{C}_2\text{H}_5}}$ II: $\overset{\text{O}}{\text{CH}_3\text{CH}_2\text{CH}_2\text{C}\text{CO}_2\text{H}}$

14.60 $\text{C}_2\text{H}_5\text{O}_2\text{CCH}_2\text{CH}_2\text{CH}_2\text{CO}_2\text{C}_2\text{H}_5$ $\xrightarrow{\text{NaOC}_2\text{H}_5}$

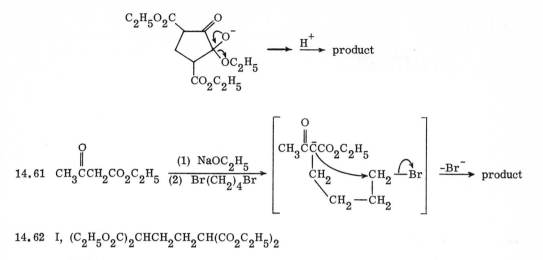

14.61 $CH_3CCH_2CO_2C_2H_5$ $\xrightarrow[\text{(2) Br}(CH_2)_4Br]{\text{(1) NaOC}_2H_5}$

14.62 I, $(C_2H_5O_2C)_2CHCH_2CH_2CH(CO_2C_2H_5)_2$

II, $C_2H_5O_2C$, $CO_2C_2H_5$, $C_2H_5O_2C$, $CO_2C_2H_5$

III, HO_2C —⟨ ⟩— CO_2H

Chapter 15
Amines

Some Important Features

Amines are compounds in which N is bonded to three other groups (or hydrogen). The nitrogen of an amine contains an unshared pair of electrons; therefore, an amine is basic and can act as a nucleophile.

$$R_3N: \; + H_2O \; \rightleftharpoons \; R_3\overset{+}{N}H \; + \; OH^-$$

$\underline{\text{an amine}}$

$$R_3N: \; + HCl \; \rightleftharpoons \; R_3\overset{+}{N}H \; Cl^-$$

$\underline{\text{an amine salt}}$

Although amines can react as nucleophiles with alkyl halides, this reaction often leads to mixtures of products.

$$R_3N: + CH_3{-}I \; \xrightarrow{S_N2} \; R_3\overset{+}{N}CH_3 \; I^-$$

$\underline{\text{only product}}$

$$RNH_2 + CH_3I \; \longrightarrow \; R\overset{+}{N}H_2CH_3 \; I^- + R\overset{+}{N}H(CH_3)_2 \; I^- + R\overset{+}{N}(CH_3)_3 \; I^-$$

For this reason, other methods for synthesizing complex amines have been developed (see Section 15.6).

Amines are <u>weak</u> bases. Their basicity is determined by the relative stabilization of amine versus amine salt.

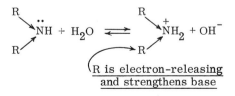

$\underline{\text{R is electron-releasing}}$
$\underline{\text{and strengthens base}}$

$$C_6H_5\overset{\cdot\cdot}{-}NH_2 + H_2O \rightleftarrows C_6H_5\overset{+}{N}H_3 + OH^-$$

resonance-stabilization
of amine weakens base

Quaternary ammonium hydroxides $\left(R_4\overset{+}{N}\ OH^-\right)$, when heated, yield Hofmann elim-ination products (the least substituted alkenes) because of steric hindrance in the transition state. (You may find it helpful to circle all β hydrogens in the quaternary ammonium hydroxide, then determine the most likely position or positions of elimination.)

OH⁻ removes this H, not this H

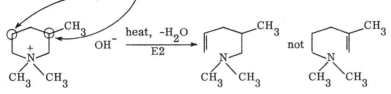

This reaction is the basis of exhaustive methylation, a reaction sequence used in structure determinations (see Section 15.13C).

Some reactions of amines (with acid halides, aldehydes, and ketones) are mentioned in Section 15.10. The formation and use of aryldiazonium salts is discussed in Section 15.12.

Reminders

In determining the relative basicities of amines, always ask yourself about the availability of the unshared electrons for donation to H^+. For example, the electrons of aniline are less available than those of an alkylamine; therefore, aniline is less basic.

If the principles behind the resolution of a racemic mixture (Section 15.9) are not clear to you, review the definitions of enantiomer and diastereomer in Chapter 4. Remember that enantiomers exhibit the same behavior toward achiral reactants or solvents, while diastereomers may exhibit different behavior.

Answers to Problems

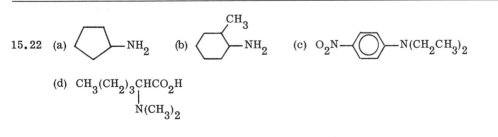

15.22 (a) [cyclopentyl]—NH_2 (b) [methylcyclohexyl]—NH_2 (c) O_2N—⟨⟩—$N(CH_2CH_3)_2$

(d) $CH_3(CH_2)_3CHCO_2H$
         |
       $N(CH_3)_2$

15.23 (a) N-propyl-2-pentylamine   (b) 1,2-cyclohexanediamine
      (c) N-ethylpiperidine   (d) 4-aminopentanal

15.24   (a) <u>N</u>-allylaniline   (b) <u>N</u>,<u>N</u>-dimethylcyclohexylamine

(c) 1-(3-chloropropyl)amine   (d) <u>N</u>-ethyl-<u>N</u>-methylbenzylamine

(e) tetraethylammonium chloride   (f) <u>N</u>,<u>N</u>-dimethylcyclohexylammonium

bromide or <u>N</u>,<u>N</u>-dimethylcyclohexylamine hydrobromide.

15.25   (a) 3°   (b) 1°   (c) salt of 3°   (d) salt of 1°   (e) quaternary   (f) 3°

(g) salt of 3°

15.26   (a) enantiomers   (b), (c), and (d) no stereoisomers

(e) two geometric isomers, each of which exists as enantiomers, for a total of

four stereoisomers

(f) enantiomers   (g) and (h) geometric isomers   (i) and (j) no stereoisomers

Compounds (g), (h), and (i) all contain $\underline{sp}^2$-hybridized nitrogen atoms. If the

groups around a C=N or N=N double bond are different, two groups can be on

the "same side" or "opposite sides." For example:

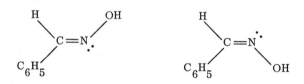

15.27   (a), (b), (c), (d), (e), (f). The remaining three compounds have no unshared elec-

trons, which are necessary for amines to act as nucleophiles.

15.28   (a) Cyclohexylamine forms stronger hydrogen bonds with water than does cyclo-

hexanol. (The partially positive H of H_2O is more strongly held by the more

basic N atom.)

(b) Dimethylamine forms hydrogen bonds, while trimethylamine does not.

(c) An ethylamine molecule, with an —NH_2 group, has <u>two</u> hydrogen atoms that

can form hydrogen bonds, while a dimethylamine molecule has only one

hydrogen atom that can form a hydrogen bond.

15.29   (a)

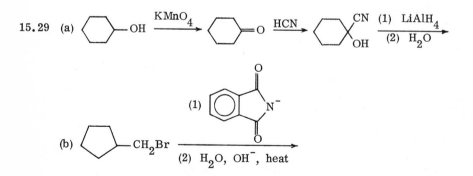

Reaction of the halide with NH_3 would not be as satisfactory as the above sequence

because of overalkylation. See Section 15.6A.

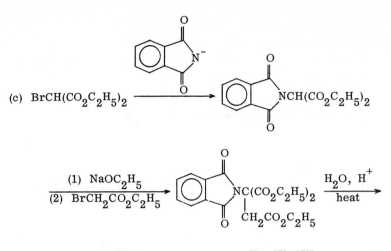

(c) $BrCH(CO_2C_2H_5)_2$

$$\xrightarrow{\text{(1) } NaOC_2H_5}{\text{(2) } BrCH_2CO_2C_2H_5}$$

$$\xrightarrow{\text{(1) } NaOC_2H_5}[\text{(2) } BrCH_2CO_2C_2H_5]$$

(d) $(CH_3)_2CHOH \xrightarrow{KMnO_4} (CH_3)_2C{=}O \xrightarrow[\text{heat, pressure}]{H_2, \ Ni, \ NH_3}$

Starting with an isopropyl halide would be unsatisfactory because of elimination side reactions.

(e) $\xrightarrow{KMnO_4}$ $\xrightarrow[\text{heat, pressure}]{H_2, \ Ni, \ H_2NCH_2CH_2CH_3}$

15.30 (a) $CH_3(CH_2)_3CH_2OH \xrightarrow{HBr} CH_3(CH_2)_3CH_2Br$

(1)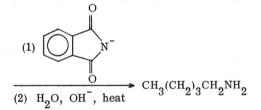

$\xrightarrow{\text{(2) } H_2O, \ OH^-, \ heat} CH_3(CH_2)_3CH_2NH_2$

(b) $[CH_3(CH_2)_3CH_2Br$ from (a)$]$

$\xrightarrow{KCN} CH_3(CH_2)_4CN \xrightarrow[\text{(2) } H_2O]{\text{(1) } LiAlH_4} CH_3(CH_2)_4CH_2NH_2$

(c) $CH_3(CH_2)_3CH_2OH \xrightarrow{KMnO_4} CH_3(CH_2)_3CO_2H \xrightarrow[\text{(2) } NH_3]{\text{(1) } SOCl_2}$

$$CH_3(CH_2)_3\overset{\overset{\displaystyle O}{\|}}{C}NH_2 \xrightarrow{Br_2, \ OH^-} CH_3(CH_2)_3NH_2$$

15.31 (a) $C_6H_6 \xrightarrow[H_2SO_4]{HNO_3} C_6H_5NO_2 \xrightarrow[HCl]{Fe} C_6H_5\overset{+}{N}H_3 \ Cl^- \xrightarrow{OH^-} C_6H_5NH_2$

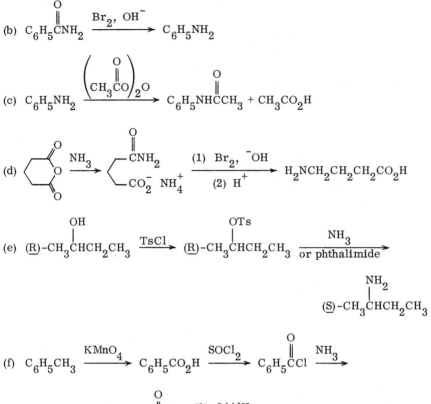

15.32 (a) Benzylamine, $C_6H_5CH_2NH_2$, is more basic because its unshared electrons are not delocalized by the aromatic pi cloud.

(b) Aniline, $C_6H_5NH_2$, is more basic because the bromine of p-bromoaniline is electron-withdrawing and decreases the electron density on the nitrogen atom.

(c) Cyclohexylamine, see (a).

(d) Tetramethylammonium hydroxide is more basic because it is more ionic, comparable to NaOH: $(CH_3)_4N^+ OH^-$.

(e) p-Nitroaniline is more basic because it has only one electron-withdrawing group.

(f) Ethylamine is more basic because ethanolamine has an electron-withdrawing group and can also form an internal hydrogen bond, both of which decrease its basicity.

(g) p-Toluidine is more basic because the —CH_3 group is electron-releasing, while the —CCl_3 group is electron-withdrawing.

15.33 In each case, the stronger base holds the proton.

(a) [pyridine ring] N + [cyclohexyl ring] NH_2^+ (b) [pyridine ring] $N + H_2O$ (c) $C_6H_5NH_2 + (CH_3)_3NH^+ \; Cl^-$

(d) no appreciable reaction

15.34 (a) no appreciable reaction because all species are completely ionized

(b) $(CH_3)_4N^+ \; {}^-O_2CCH_3 + H_2O$ (c) [cyclohexyl ring] $NH_2^+ \; {}^-O_2CCH_3$

(d) [phthalimide ring structure with two C=O groups and N$^-$] $N^- + CH_3OH$

15.35 (a) (3), (1), (2)   (b) (2), (3), (1)   (c) (1), (2)   (d) (3), (2), (1)
(e) (1), (2), (3)   (f) (2), (1)

15.36 (a) Calculate $\underline{K_b}$. $\underline{K_b} = 10^{-3.34}$

$$= 10^{0.66} \times 10^{-4}$$

$$= 4.57 \times 10^{-4}$$

Substitute:
$$\frac{\left[CH_3\overset{+}{N}H_3\right]\left[OH^-\right]}{[CH_3NH_2]} = \underline{K_b}$$

$$\frac{\left[CH_3\overset{+}{N}H_3\right](0.0100)}{(0.00100)} = 4.57 \times 10^{-4}$$

$$\left[CH_3\overset{+}{N}H_3\right] = \frac{(4.57 \times 10^{-4})(10^{-3})}{(10^{-2})}$$

$$= 4.57 \times 10^{-5} \; \underline{M}$$

(b) $\dfrac{x[OH^-]}{\underline{x}} = 4.57 \times 10^{-4}$

$$[OH^-] = 4.57 \times 10^{-4} \; \underline{M}$$

$$pOH = 3.34$$

$$pH + pOH = 14 \quad \text{(definition)}$$

$$pH = 14 - 3.34 = 10.7$$

15.37 (a) Dissolve in diethyl ether. Wash with dilute acid to remove the amine. Wash with dilute $NaHCO_3$ to remove the acid. The alcohol remains in the ether.

(b) Washing with dilute acid removes the amine.

15.38 The nitrogen of the NCH_3 group is the most basic. The other nitrogens are part of an amide group (not basic) and part of an aromatic ring system (not basic in this case). Even though we have not yet discussed aromatic heterocycles, you should have predicted a very low basicity for this latter N because of delocalization of the unshared electrons.

15.39 (a) $(CH_3CH_2)_2\overset{+}{N}HCH_2CH_2O_2C$—⬡—$NH_2$ HSO_4^-

(b) $(CH_3CH_2)_2\overset{+}{N}HCH_2CH_2OH$ Cl^- + HO_2C—⬡—$\overset{+}{N}H_3$ Cl^-

(c) $(CH_3CH_2)_2NCH_2CH_2OH$ + Na^+ ^-O_2C—⬡—NH_2

15.40 (a) $(CH_3CH_2)_2\overset{+}{N}HCH_2\overset{\overset{O}{\|}}{C}NH$—⬡$\begin{smallmatrix}CH_3\\CH_3\end{smallmatrix}$ HSO_4^-

(b) $(CH_3CH_2)_2\overset{+}{N}HCH_2CO_2H$ Cl^- + $H_3\overset{+}{N}$—⬡$\begin{smallmatrix}CH_3\\CH_3\end{smallmatrix}$ Cl^-

(c) $(CH_3CH_2)_2NCH_2CO_2^-$ Na^+ + H_2N—⬡$\begin{smallmatrix}CH_3\\CH_3\end{smallmatrix}$

(The hindrance around the amide group might prevent reaction in b and c.)

15.41 (a) Treat with one enantiomer of a chiral carboxylic acid, separate the diastereomeric salts, and regenerate the amine with dilute base.

(b) Saponify the ester with dilute NaOH, acidify, treat the racemic carboxylic acid with one enantiomer of a chiral amine, separate the diastereomeric salts, make each salt alkaline, remove the amine by extraction with an organic solvent, and acidify the aqueous solution to regenerate the acid. To regenerate the ester, treat the acid with CH_2N_2. (Alternatively, treat the carboxylate with CH_3I, or treat the acid with CH_3OH + HCl.)

15.42 (a) $CH_3\overset{+}{N}H_3\ I^-$ (b) $(CH_3)_4\overset{+}{N}\ I^-$ (c) $NH_3 + {}^-O_2CCH_2CH_2CO_2^-$

(d) $CH_3NH_2 + {}^-O_2CCH_2CH_2CO_2^-$

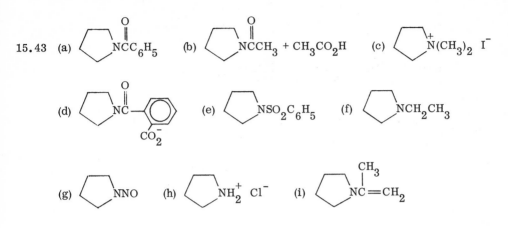

15.43 (a), (b), (c), (d), (e), (f), (g), (h), (i)

15.44 No reaction for (a), (b), (d), (e), (f), and (i).

15.45 (a) Treat with cold HNO_2. Aniline forms a diazonium salt, while n-hexylamine gives off nitrogen gas.

(b) Treat with cold, dilute HCl. n-Octylamine dissolves, while octanamide does not.

(c) Treat with dilute NaOH:

$$(CH_3CH_2)_3\overset{+}{N}H\ Cl^- \xrightarrow{\ OH^-\ } (CH_3CH_2)_3N + H_2O + Cl^-$$
amine smell

$$(CH_3CH_2)_4\overset{+}{N}\ Cl^- \xrightarrow{\ OH^-\ } \text{no reaction (odorless)}$$

15.46 (a), (b), (c), (d)

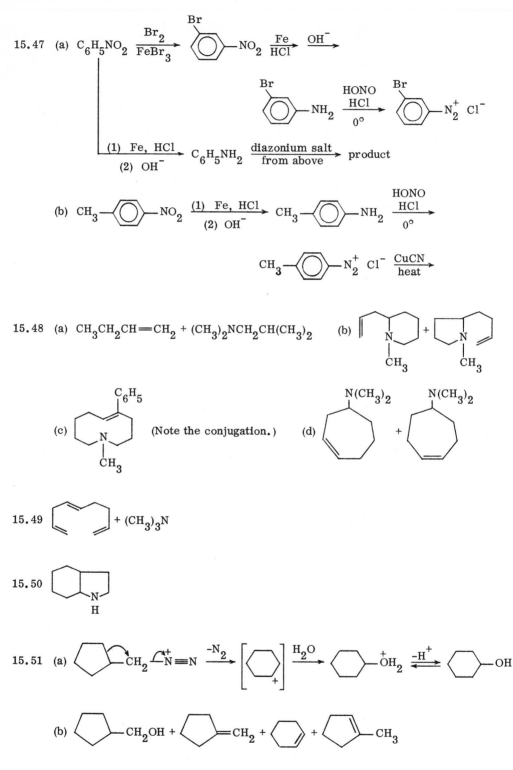

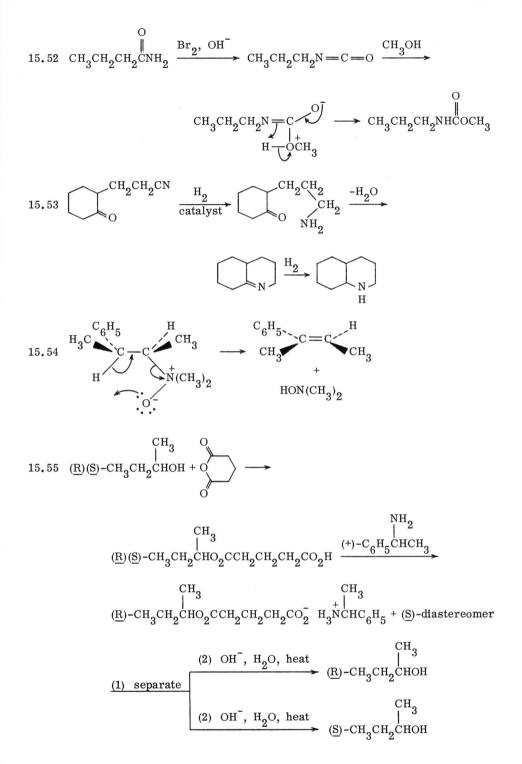

15.52   CH₃CH₂CH₂CNH₂ $\xrightarrow{\text{Br}_2,\ \text{OH}^-}$ CH₃CH₂CH₂N=C=O $\xrightarrow{\text{CH}_3\text{OH}}$

CH₃CH₂CH₂N=C⟨Oⁿ⟩ → CH₃CH₂CH₂NHCOCH₃
H—OCH₃

15.53

H₂ / catalyst   →   -H₂O →

H₂ →

15.54

C₆H₅—C=C—H + HON(CH₃)₂

15.55   (R)(S)-CH₃CH₂CHOH + →

(R)(S)-CH₃CH₂CHO₂CCH₂CH₂CH₂CO₂H $\xrightarrow{(+)\text{-}C_6H_5\text{CHCH}_3}$

(R)-CH₃CH₂CHO₂CCH₂CH₂CH₂CO₂⁻ H₃N⁺CHC₆H₅ + (S)-diastereomer

(1) separate

(2) OH⁻, H₂O, heat → (R)-CH₃CH₂CHOH

(2) OH⁻, H₂O, heat → (S)-CH₃CH₂CHOH

15.56 $\underline{A}$: $CH_3CH_2NH_2$ $\underline{B}$: $(CH_3CH_2)_2NH$ $\underline{C}$: $CH_3CH_2NH\overset{\displaystyle O}{\overset{\displaystyle \|}{C}}CH_3$

$\underline{D}$: $(CH_3CH_2)_2N\overset{\displaystyle O}{\overset{\displaystyle \|}{C}}CH_3$

15.57 (a) $C_6H_5CH_2CH_2NH_2$ and (b) $C_6H_5\overset{\displaystyle CH_3}{\overset{\displaystyle |}{C}}HNH_2$

Besides nmr absorption from aryl protons, (a) would exhibit two triplets and a
singlet (area ratio, 1:1:1). However, (b) would show a quartet, a doublet, and a
singlet (area ratio, 1:3:2).

15.58 A: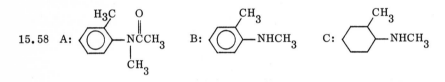

Chapter 16

Polycyclic and Heterocyclic Aromatic Compounds

Some Important Features

The polycyclic aromatic compounds, such as naphthalene and anthracene, are not symmetrical as benzene is. Consequently, some carbon–carbon bonds of the polycyclic aromatic compounds have more double-bond character than others.

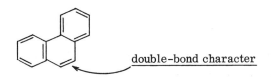

double-bond character

The aromatic polycyclic compounds are more reactive toward electrophiles, oxidizing agents, and reducing agents than is benzene because the intermediates still contain one or more rings with aromatic character.

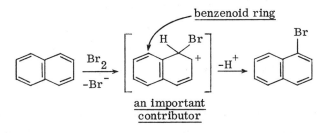

benzenoid ring

an important
contributor

Sulfonation of naphthalene, like the sulfonation of benzene (Section 10.9F), is reversible. Thus, the sulfonation of naphthalene can lead to substitution at the 1- or 2-position, depending on the reaction conditions (see Section 16.5A).

The position of the second substitution on naphthalene is determined by the first substituent.

activating

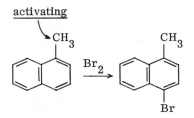

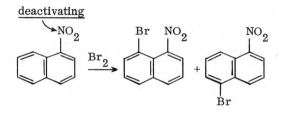

deactivating

Although pyridine is a weaker base than a tertiary amine (because the N is sp^2 hybridized), pyridine still reacts with acids or with alkyl halides to yield salts.

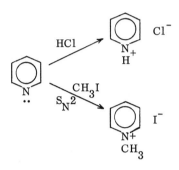

Compared to benzene, pyridine is deactivated to electrophilic substitution and activated to nucleophilic substitution because the nitrogen withdraws electron density from the rest of the ring.

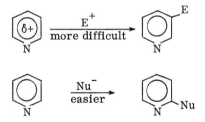

Pyrrole, unlike pyridine, is not basic because it has no unshared electrons; the "extra" pair of electrons on the nitrogen are both contributed toward the aromatic pi cloud.

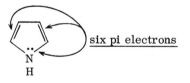

six pi electrons

Because the nitrogen is electron-deficient, the rest of the pyrrole ring is electron-rich and undergoes electrophilic substitution more easily than benzene.

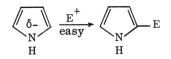

Other important features of this chapter are quinoline and isoquinoline (Section 16.8), alkaloids (16.10), porphyrins (16.11), nucleic acids and the importance of hydrogen bonding in these compounds (16.12), and the nucleotides ATP, NAD^+, and FAD (16.13).

Reminders

Review Chapter 10 if aromaticity and aromatic substitution reactions are not clear to you. Review Chapter 15 if the structural features controlling the relative basicities of amines are not clear.

Electrophilic substitution occurs under acidic conditions (or neutral conditions for activated rings). Keep in mind the relative reactivities of aromatic compounds toward electrophilic substitution.

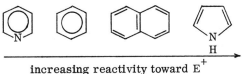

increasing reactivity toward E^+

Nucleophilic substitution occurs only with a strong base under any circumstances.

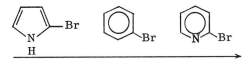

increasing reactivity toward Nu^-

Answers to Problems

16.13 (a) 2-propylnaphthalene   (b) 2-propylpyridine   (c) 1-phenylnaphthalene
     (d) 2,3-dimethylfuran   (e) 4-methylquinoline

16.14 (a) 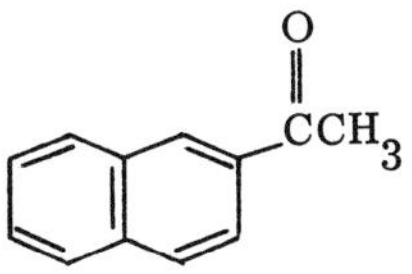 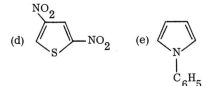   (b) 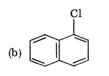   (c)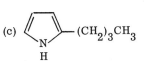

     (d) 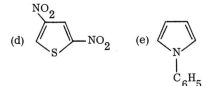   (e)

16.15 (a) 14, aromatic   (b) 12   (c) 14, aromatic   (d) 16   (e) 12   (f) 8

(g) 10, aromatic

We can tell by inspection that the entire ring systems in compounds (b), (e), and (f) are not aromatic because not all the carbon atoms are sp^2 hybridized. Compound (g) is aromatic because the nitrogen atom contributes its two unshared electrons toward the aromatic pi cloud. Compound (d) appears at first glance to be fully aromatic, but it does not contain $4\underline{n} + 2$ pi electrons. It is thought that an aromatic pi cloud encompasses the peripheral carbon atoms in this structure.

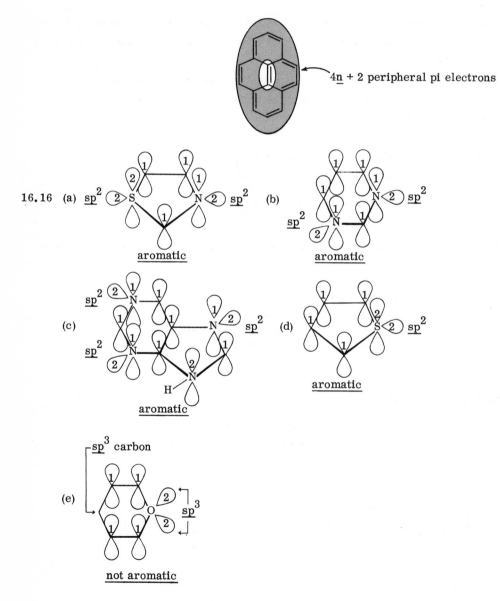

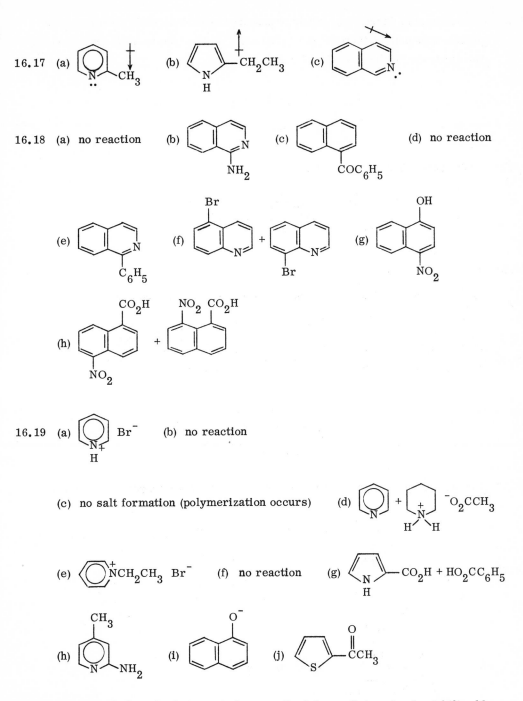

16.17 (a) ... (b) ... (c) ...

16.18 (a) no reaction   (b) ...   (c) ...   (d) no reaction

(e) ...   (f) ... + ...   (g) ...

(h) ... + ...

16.19 (a) ... Br⁻   (b) no reaction

(c) no salt formation (polymerization occurs)   (d) ... + ... ⁻O₂CCH₃

(e) ... Br⁻   (f) no reaction   (g) ... —CO₂H + HO₂CC₆H₅

(h) ...   (i) ...   (j) ...

16.20 Reaction (b) has the faster rate because the intermediate anion is stabilized by electron-withdrawal by the chlorine.

16.21 The —CO₂H group is electron-withdrawing and deactivates the ring toward electrophilic substitution.

16.22   The nitrogen in thiazole has a pair of unshared electrons, while the nitrogen in pyrrole does not. In thiazole, the sulfur atom, rather than the nitrogen, provides two electrons for the aromatic pi cloud.

16.23   The negative charge is delocalized:

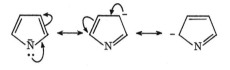

16.24   (b), (c), (a), (d)

(b) is less reactive than (c) because the active positions (2 and 5) are blocked.

16.25

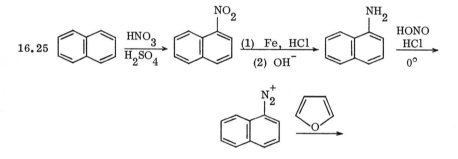

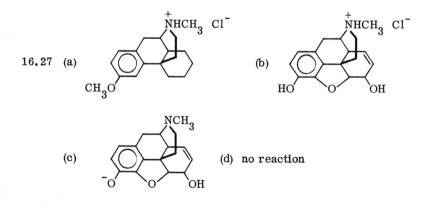

16.26   The more basic nitrogen is the alkylamine N; the other nitrogen is an amide N.

16.27   (a)

(b)

(c)

(d) no reaction

16.28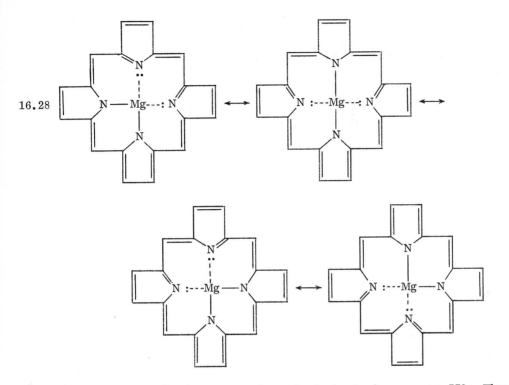

16.29 The hydrogen bonding between guanine and cytosine is shown on page 772. There
are three hydrogen bonds between the two. Adenine and cytosine can form only one
hydrogen bond.

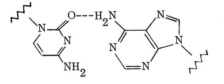

16.30 (a)

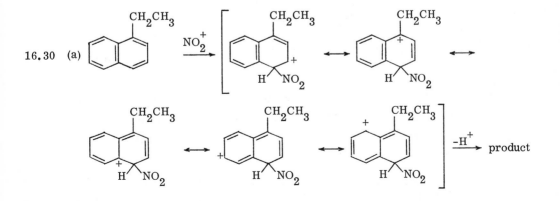

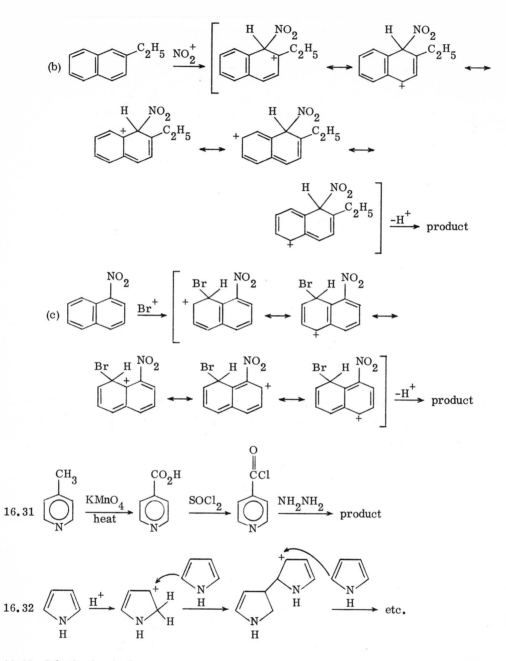

(b)

(c)

16.31

16.32

16.33 Substitution in the 3-position is favored because (1) the nitrogen ring is electron-rich, and (2) the intermediate has two resonance structures with the aromatic benzene ring intact (compared to one such structure   for 2-substitution).

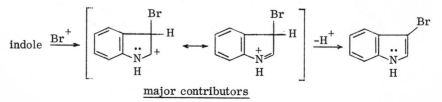

major contributors

16.34   Pyrimidine is less reactive toward electrophilic attack and more reactive toward
nucleophilic attack because the two nitrogens decrease the electron density of the
ring carbons to a greater extent.

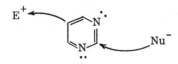

16.35   Imidazole and the imidizolium ion each has six electrons in an aromatic pi cloud,
as the p-orbital pictures show.

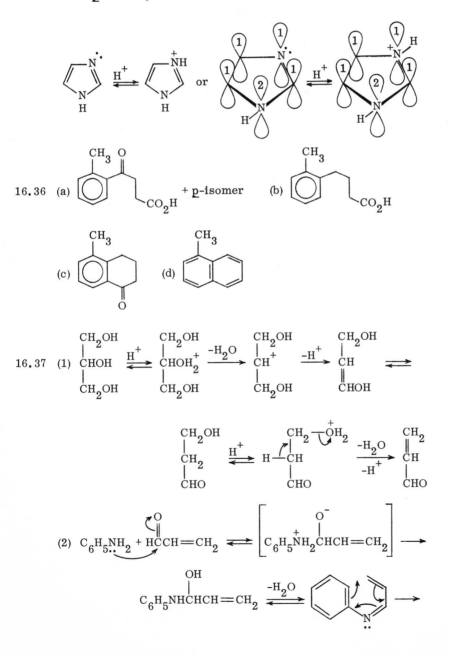

16.36   (a) [structure: CH₃-substituted benzene ring with C(=O)CH₂CH₂CO₂H] + p-isomer   (b) [structure: CH₃-substituted benzene ring with CH₂CH₂CO₂H]

(c) [structure: CH₃-substituted tetralone with O] (d) [structure: CH₃-substituted naphthalene]

16.37   (1) [reaction sequence]

(2) $C_6H_5NH_2$ + [structure] ...

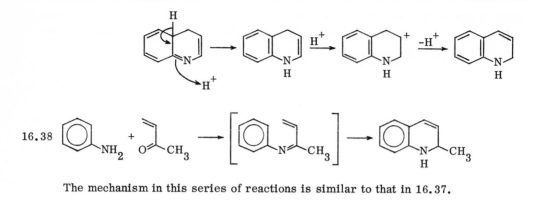

16.38

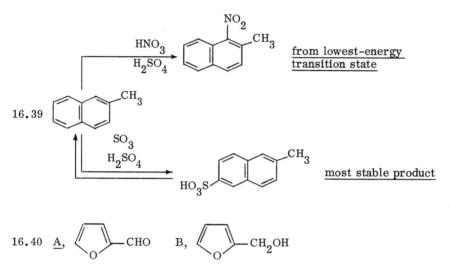

The mechanism in this series of reactions is similar to that in 16.37.

16.39

from lowest-energy transition state

most stable product

16.40 <u>A</u>,   —CHO   B,   —CH$_2$OH

Chapter 17

Carbohydrates

Some Important Features

A Fischer projection is often used to show the structure and configuration of a monosaccharide.

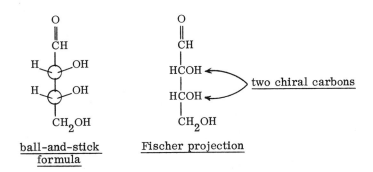

ball-and-stick formula   Fischer projection

Glucose and the other monosaccharides exist primarily as pairs of cyclic hemi-acetals (<u>anomers</u>), which are in equilibrium with the open-chain carbonyl form in solution. Hemiacetals can undergo aldehyde reactions because of this equilibrium.

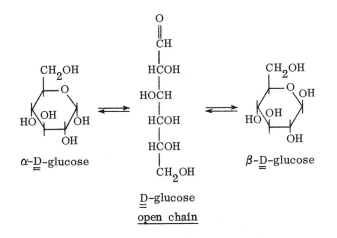

The monosaccharides form acetals (<u>glycosides</u>) when treated with an alcohol. A glycoside is not in equilibrium with the carbonyl form in neutral or alkaline solution;

therefore, glycosides do not undergo aldehyde reactions. Glycoside links can be hydrolyzed in acidic solution or with appropriate enzymes.

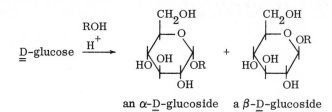

an α-D-glucoside   a β-D-glucoside

The relative configurations of sugars may be determined by synthesis.

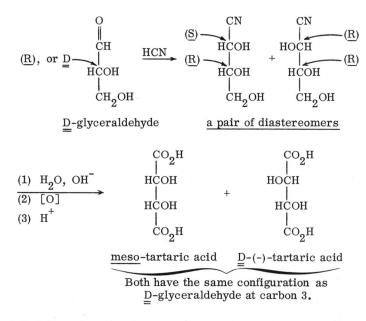

D-glyceraldehyde   a pair of diastereomers

meso-tartaric acid   D-(-)-tartaric acid

Both have the same configuration as
D-glyceraldehyde at carbon 3.

The original determination of the structures of some monosaccharides is discussed
in Section 17.9.

Some important reactions of monosaccharides are the oxidation to aldonic acids
(carbon 1 oxidized), aldaric acids (both carbon 1 and the last carbon oxidized), and uronic
acids (only the last carbon oxidized). These oxidations are discussed in Section 17.6. The
monosaccharides also can be reduced to alditols (Section 17.7). The hydroxyl groups can
be esterified or converted to alkoxyl groups (Section 17.8).

Disaccharides are formed from two monosaccharide units joined by a glycoside link.

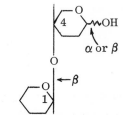

α or β

β

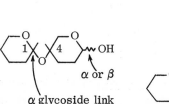

α or β

α glycoside link

Polysaccharides are composed of many monosaccharide units joined by glycoside links. Partial structures of some important polysaccharides are shown in Section 17.11.

Reminders

Memorize the Fischer projections and Haworth formulas for α- and β-$\underline{\underline{D}}$-glucose. Then, interconversions for other sugars are simplified.

If you forget the configurations of the carbon atoms of glucose, draw the favored chair form. β-$\underline{\underline{D}}$-Glucose has all substituents in equatorial positions.

A review of Chapter 4 would be valuable. Be sure you know what a <u>meso</u> compound is and how to assign ($\underline{R}$) and ($\underline{S}$) configurations.

Answers to Problems

17.21 (a) 4   (b) (3)   (c) (1)   (d) (2)

Note that an -<u>ose</u> ending refers to an aldehyde or hemiacetal (or ketone or hemiketal), while an -<u>oside</u> ending refers to a glycoside (an acetal or ketal).

17.22 all $\underline{\underline{D}}$

17.23 (1)

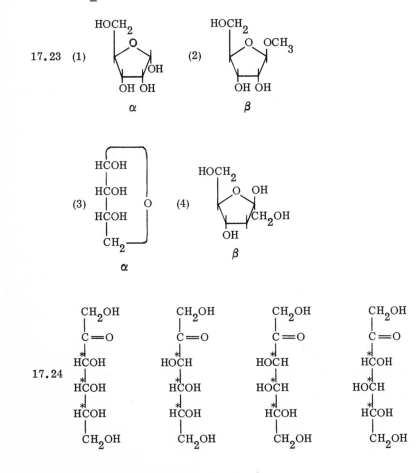

17.24

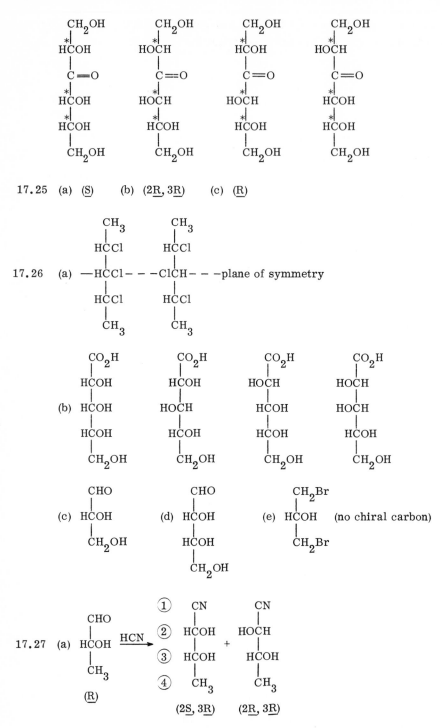

17.25   (a) (S)   (b) (2R, 3R)   (c) (R)

17.26   (a) —HCCl— — —ClCH— — —plane of symmetry

(b)

(c) CHO
     |
     HCOH
     |
     CH₂OH

(d) CHO
     |
     HCOH
     |
     HCOH
     |
     CH₂OH

(e) CH₂Br
     |
     HCOH   (no chiral carbon)
     |
     CH₂Br

17.27   (a)

(b) Yes. Although carbon 2 in the product is an equal mixture of (R) and (S), carbon 3 is (R) in each case.

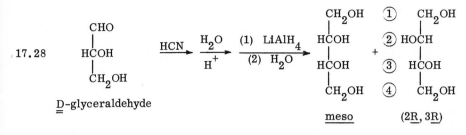

17.28

D-glyceraldehyde

The product mixture is a mixture of diastereomers and thus can be separated by physical means. Each isomer can then be compared with the unknown tetraol. The unknown tetraol is either meso (has no optical rotation); (2R, 3R) (same rotation as the tetraol from D-glyceraldehyde); or (2S, 3S) (opposite rotation).

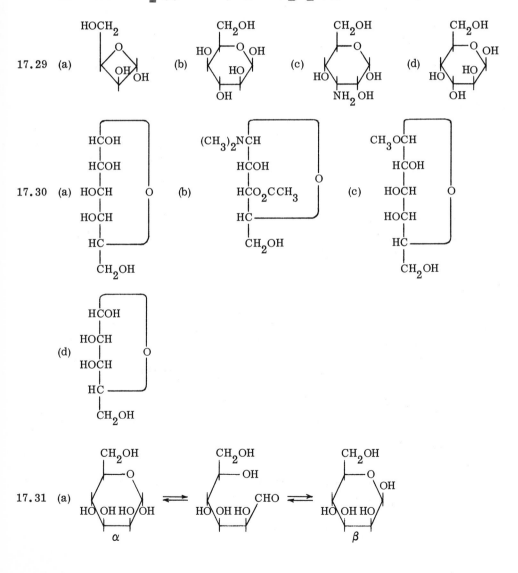

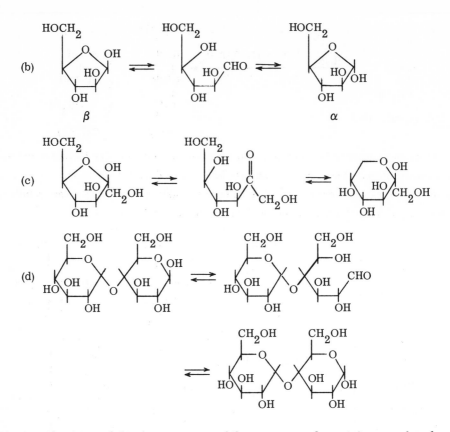

17.32   (a), (d), (e), and (h), because none of these compounds contains a carbonyl or hemiacetal group; all are glycosides, which are not in equilibrium with the carbonyl compounds or the anomers.

17.33   (a), (d), (e), (h)

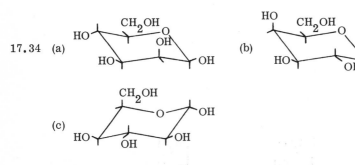

In each case, the most stable conformation is the one in which the greatest number of groups is equatorial.

17.35 In β-D-glucopyranose, all substituents can be equatorial; therefore, it is the most stable of all the aldopyranoses.

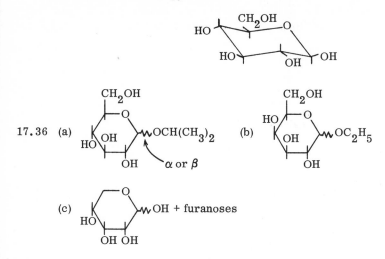

17.36 (a) ... WOCH(CH₃)₂ α or β   (b) ... WOC₂H₅

(c) ... WOH + furanoses

17.37 Tollens reagent contains $Ag(NH_3)_2^+$ and OH^- ions. In base, D-fructose can be converted to D-glucose and D-mannose by way of an enediol intermediate. (These two sugars differ in configuration only at carbon 2.) Oxidation of the aldehyde groups in these sugars therefore yields aldonic acids of both sugars.

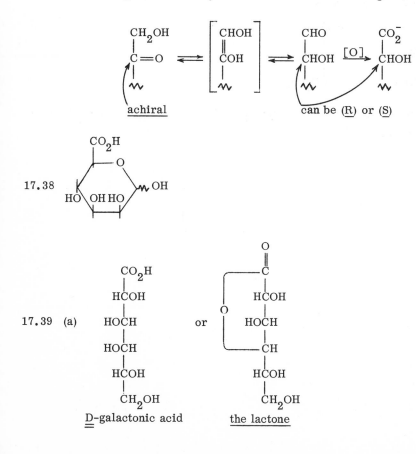

17.38

17.39 (a)   CO₂H | HCOH | HOCH | HOCH | HCOH | CH₂OH
D-galactonic acid   or   the lactone

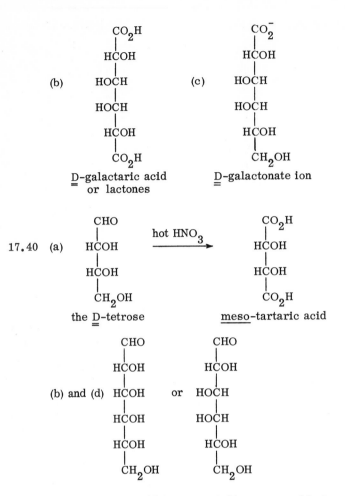

(b)

$$CO_2H$$
$$HCOH$$
$$HOCH$$
$$HOCH$$
$$HCOH$$
$$CO_2H$$

D-galactaric acid
or lactones

(c)

$$CO_2^-$$
$$HCOH$$
$$HOCH$$
$$HOCH$$
$$HCOH$$
$$CH_2OH$$

D-galactonate ion

17.40 (a)

$$CHO$$
$$HCOH$$
$$HCOH$$
$$CH_2OH$$

the D-tetrose

$$\xrightarrow{\text{hot HNO}_3}$$

$$CO_2H$$
$$HCOH$$
$$HCOH$$
$$CO_2H$$

meso-tartaric acid

(b) and (d)

$$CHO$$
$$HCOH$$
$$HCOH$$
$$HCOH$$
$$HCOH$$
$$CH_2OH$$

or

$$CHO$$
$$HCOH$$
$$HOCH$$
$$HOCH$$
$$HCOH$$
$$CH_2OH$$

Each of these D-aldohexoses yields a meso-aldaric acid when oxidized and a meso-alditol when reduced.

(c)

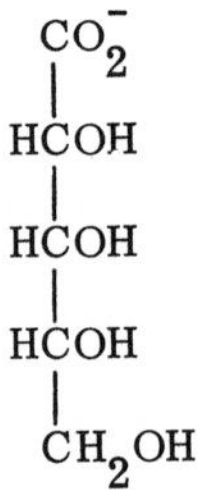

$$CO_2^-$$
$$HCOH$$
$$HCOH$$
$$HCOH$$
$$CH_2OH$$

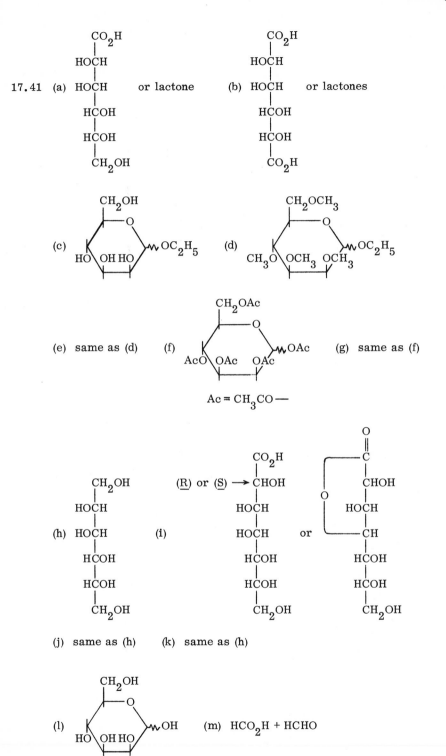

17.41 (a) CO_2H — $HOCH$ — $HOCH$ — $HCOH$ — $HCOH$ — CH_2OH or lactone

(b) CO_2H — $HOCH$ — $HOCH$ — $HCOH$ — $HCOH$ — CO_2H or lactones

(c)   [ring structure: CH_2OH, O, HO, OH, HO, $\sim\sim OC_2H_5$]

(d)   [ring structure: CH_2OCH_3, O, CH_3O, OCH_3, OCH_3, $\sim\sim OC_2H_5$]

(e)   same as (d)

(f)   [ring structure: CH_2OAc, O, AcO, OAc, OAc, $\sim\sim OAc$]

$Ac = CH_3CO-$

(g)   same as (f)

(h) CH_2OH — $HOCH$ — $HOCH$ — $HCOH$ — $HCOH$ — CH_2OH

(i)   (R) or (S) → CO_2H — $CHOH$ — $HOCH$ — $HOCH$ — $HCOH$ — $HCOH$ — CH_2OH or   [lactone structure: O, C=O, $CHOH$, $HOCH$, CH, $HCOH$, $HCOH$, CH_2OH]

(j)   same as (h)

(k)   same as (h)

(l)   [ring structure: CH_2OH, O, HO, OH, HO, $\sim\sim OH$]

(m) $HCO_2H + HCHO$

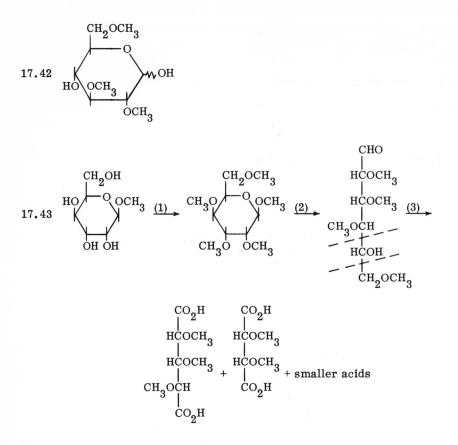

17.42

17.43   (1)→   (2)→   (3)→

+ smaller acids

17.44 (a) maltose, positive Tollens test; sucrose, negative.

(b) Treat with HNO_3, heat; $\underline{D}$-xylose → $\underline{meso}$-diacid and $\underline{D}$-lyxose → optically active diacid.

17.45 eleven: α,α-1,1′; α,β-1,1′; β,β-1,1′; α-1,2′; β-1,2′; α-1,3′; β-1,3′; α-1,4′; β-1,4′; α-1,6′; β-1,6′

17.46 maltose and

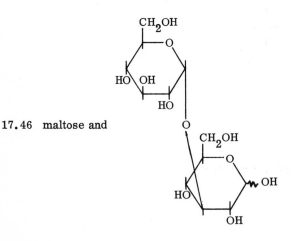

17.47 α, α'-, α, β'-, or β, β'-1, 1'-$\underline{\underline{D}}$-glycopyranoside.   (Because trehalose is nonreducing, we know that carbon 1 of the first unit is joined to carbon 1 of the second unit.)

17.48   A,

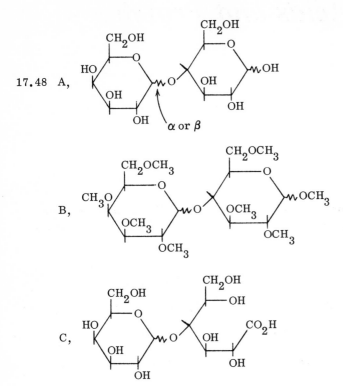

B,

C,

Chapter 18

Amino Acids and Proteins

Some Important Features

Proteins are polyamides formed from L-α-amino acids. These amino acids may be acidic (acidic side chain), basic (basic side chain), or neutral (side chain neither acidic nor basic).

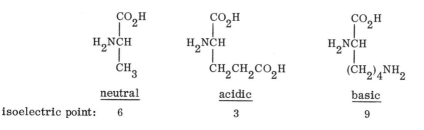

	neutral	acidic	basic
isoelectric point:	6	3	9

Amino acids exist as dipolar ions:

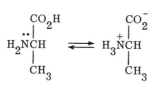

Some techniques for synthesizing amino acids are discussed in Section 18.4.

A small protein molecule is called a peptide. The synthesis of peptides is accomplished by blocking some reactive groups within the amino acids and then allowing other functional groups to react (Section 18.8). In biological systems, the joining of amino acids into protein chains is accomplished enzymatically by mRNA, ribosomes, and tRNA (Section 18.9).

The order of attachment of amino acids (the primary structure) in peptides and proteins can be determined by partial hydrolysis and terminal residue analysis (Section 18.7).

Hydrogen bonding between NH and C$=$O groups and side chain interactions allow a protein to assume a secondary structure (the shape of a chain) and possibly a tertiary structure (interactions between different parts of a chain or between two or more chains).

Thus, a chain can form a helix that can fold into a globule or that can interact with other helices. The secondary and tertiary structures lend strength or solubility to a protein. When these higher structures are disrupted by a change in environment (such as a change in pH), the protein is said to be <u>denatured</u>.

Enzymes are proteins that act as biological catalysts. Their specific catalytic action depends upon the unique protein surface presented to substrates and upon the active site, which may be a <u>coenzyme</u> (a nonprotein organic molecule or a metal ion). Many vitamins are coenzymes.

Reminders

In predicting acid-base reactions of amino acids or peptides, look for the most acidic and most basic sites in the molecule.

Remember that any synthesis from an achiral molecule leads to an achiral or racemic product; however, a racemic mixture of enantiomers can be separated by procedures outlined in Chapter 15.

Answers to Problems

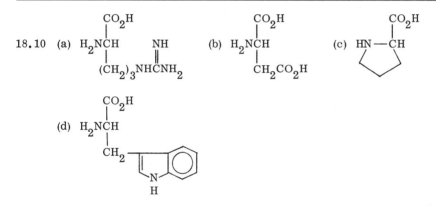

18.11 (a) neutral   (b) acidic   (c) neutral   (d) neutral   (e) basic   (f) neutral

18.12 $SCH_2\overset{O}{\overset{\|}{C}}NHCHCO_2H$ with CH_2SH; SCH_2CHCO_2H, NH_2

$HSCH_2\overset{O}{\overset{\|}{C}}NHCHCO_2H$; $CH_2S-SCH_2CHCO_2H$, NH_2

18.13 (a) CH_3CHCO_2H, $\overset{+}{N}H_3$ Cl^- (b) $CH_3CHCO_2^-$ K^+, NH_2 (c) $CH_3CHCO_2CH_3$, $\overset{+}{N}H_3$ HSO_4^-

(d) $CH_3CHCO_2H + CH_3CO_2H$
      |
      $NHCCH_3$
         $\|$
         O

18.14 $(CH_3)_2CHCH_2CO_2H$ $\xrightarrow[PBr_3]{Br_2}$ $(CH_3)_2CHCHBrCO_2H$ $\xrightarrow{\text{excess } NH_3}$

     achiral   racemic

                       NH_2 NH_2
                        |
         $(CH_3)_2CHCHCO_2^-$ NH_4^+ $\xrightarrow{H^+}$ racemic $(CH_3)_2CHCHCO_2H$

             racemic

       CO_2H CO_2H CO_2H
        | hot $KMnO_4$ | H_2, NH_3 |
18.15 (R)–HCOH $\xrightarrow{\text{hot } KMnO_4}$ $C=O$ $\xrightarrow[Pd]{H_2,\ NH_3}$ racemic $CHNH_2$
        | | |
        CH_3 CH_3 CH_3

                              achiral

18.16 (a) An amino acid contains a carboxylate group $\left(-CO_2^-\right)$, rather than a carboxyl group $(-CO_2H)$.

     (b) When the solution is acidified, the carboxyl group is generated.

                                     NH_2 NH_2
                      NH_3 | H_2O, H^+ |
18.17 (a) $C_6H_5CH_2CHO$ $\xrightarrow{NH_3 / HCN}$ $C_6H_5CH_2CHCN$ $\xrightarrow{H_2O,\ H^+}$ $C_6H_5CH_2CHCO_2H$

                                 NH_2 NH_2
                      NH_3 | H_2O, H^+ |
     (b) $(CH_3)_2CHCHO$ $\xrightarrow{NH_3 / HCN}$ $(CH_3)_2CHCHCN$ $\xrightarrow{H_2O,\ H^+}$ $(CH_3)_2CHCHCO_2H$

     (c) both racemic

18.18 (a) $CH_3CHCO_2^-\ \overset{+}{N}H_4$ (b) $Cl^-\ \overset{+}{N}H_3CH_2CO_2H +$
               |
               NH_2

                                                             (a benzene ring with two CO_2H groups ortho)

     (c) $C_6H_5CH_2CHCO_2H +$ (a benzene ring with two CO_2H groups ortho)   (d) $(CH_3)_2CHCH_2CHCN$
                   $\overset{+}{N}H_3\ Cl^-$ NH_2

     (e) $C_6H_5CH_2CHCO_2CH_3$
                 $\overset{+}{N}H_3\ Cl^-$

18.19 (a) a neutral amino acid: (3)   (b) an acidic amino acid: (4)

(c) a neutral amino acid: (2)   (d) a basic amino acid: (1)

Note that cysteine (a) is more acidic than proline (c). The reason is that proline contains a more basic 2° amino group.

18.20 (a) neutral, 6   (b) slightly basic, 8 (See Answer 18.41)   (c) acidic, 3

(d) neutral, 6   (e) basic, 10

18.21 (a) $\overset{+}{H_3N}CHCO_2^-$ + Cl^- (b) $\overset{+}{H_3N}CHCO_2^-$ + Cl^-

$\quad\quad\quad$ $CH(CH_3)_2$ $\quad\quad\quad\quad\quad\quad$ $(CH_2)_4NH_2$

(c) $\overset{+}{H_3N}CHCO_2^-$ + Na^+ + H_2O (d) $\overset{+}{H_3N}CHCO_2^-$ + Na^+ + H_2O

$\quad\quad\quad$ $CH(CH_3)_2$ $\quad\quad\quad\quad\quad\quad$ $(CH_2)_4\overset{+}{NH_3}$

In each case, look for the more acidic or more basic group in the reactant molecule.

18.22 (a)

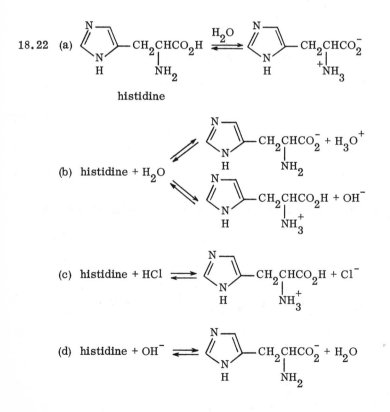

histidine

(b) histidine + H_2O

(c) histidine + HCl

(d) histidine + OH^-

18.23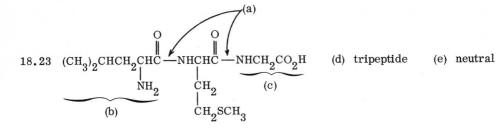

(d) tripeptide   (e) neutral

18.24 (a) dipeptide   (b) and (c) tripeptides   (d) decapeptide

18.25 (a) $NH_2CH_2\overset{\displaystyle O}{\overset{\|}{C}}$—$NHCH_2CO_2H$ (b) $NH_2\underset{\underset{\displaystyle CH_3}{|}}{CH}\overset{\displaystyle O}{\overset{\|}{C}}$—$NH\underset{\underset{\underset{\underset{\displaystyle CH(CH_3)_2}{|}}{CH_2}}{|}}{CH}\overset{\displaystyle O}{\overset{\|}{C}}$—$NHCHCO_2H$
$\underset{\displaystyle CH_2CH_2SCH_3}{|}$

18.26 (a) serylglycylleucine, ser-gly-leu

(b) prolylthreonylmethionine, pro-thr-met

18.27 (a) $H_3\overset{+}{N}CH_2\overset{\displaystyle O}{\overset{\|}{C}}$—$NH\underset{\underset{\displaystyle (CH_2)_4NH_3^+}{|}}{CH}CO_2H$ (b) $H_2NCH_2\overset{\displaystyle O}{\overset{\|}{C}}NH\underset{\underset{\displaystyle CH_2CH_2CO_2^-}{|}}{CH}CO_2^-$

(c) $H_2NCH_2\overset{\displaystyle O}{\overset{\|}{C}}NHCHCO_2^-$
$\underset{\displaystyle CH_2}{|}$—〈benzene ring〉—$O^-$

18.28 The protein contains either an unsubstituted amide group, $-\overset{\displaystyle O}{\underset{\displaystyle O}{\overset{\|}{C}}}NH_2$, or arginine.

18.29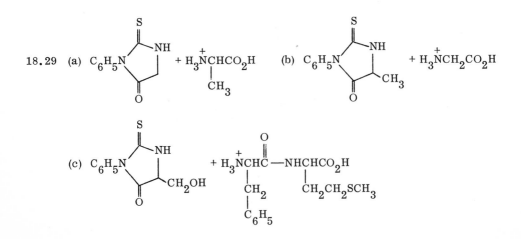

18.30 (a) and (b), $H_2NCH_2CO_2H + H_2NCHCO_2H$

                                                           $\overset{|}{CH_3}$

(c) $H_2NCH\overset{O}{\overset{\|}{C}}-NHCHCO_2H + H_2NCHCO_2H$

      $\overset{|}{HOCH_2} \qquad \overset{|}{CH_2C_6H_5} \qquad \overset{|}{CH_2CH_2SCH_3}$

18.31 arg-pro-pro-gly-phe-ser-pro-phe-arg

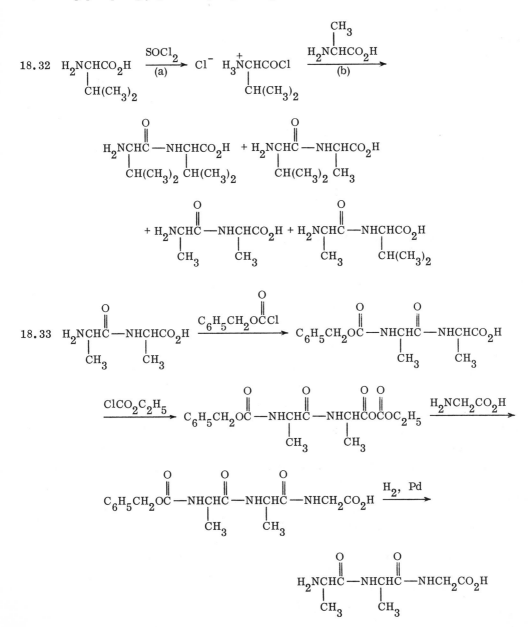

18.32 $H_2NCHCO_2H \xrightarrow[\text{(a)}]{SOCl_2} Cl^- \ H_3\overset{+}{N}CHCOCl \xrightarrow[\text{(b)}]{\overset{CH_3}{\overset{|}{H_2NCHCO_2H}}}$

        $\overset{|}{CH(CH_3)_2}$ $\overset{|}{CH(CH_3)_2}$

$H_2NCH\overset{O}{\overset{\|}{C}}-NHCHCO_2H + H_2NCH\overset{O}{\overset{\|}{C}}-NHCHCO_2H$

 $\overset{|}{CH(CH_3)_2} \overset{|}{CH(CH_3)_2} \qquad \overset{|}{CH(CH_3)_2} \overset{|}{CH_3}$

$+ H_2NCH\overset{O}{\overset{\|}{C}}-NHCHCO_2H + H_2NCH\overset{O}{\overset{\|}{C}}-NHCHCO_2H$

   $\overset{|}{CH_3} \qquad \overset{|}{CH_3} \qquad\quad \overset{|}{CH_3} \quad \overset{|}{CH(CH_3)_2}$

18.33 $H_2NCH\overset{O}{\overset{\|}{C}}-NHCHCO_2H \xrightarrow{C_6H_5CH_2O\overset{O}{\overset{\|}{C}}Cl} C_6H_5CH_2O\overset{O}{\overset{\|}{C}}-NHCH\overset{O}{\overset{\|}{C}}-NHCHCO_2H$

 $\overset{|}{CH_3} \quad \overset{|}{CH_3} \qquad\qquad\qquad\qquad\qquad \overset{|}{CH_3} \quad\ \overset{|}{CH_3}$

$\xrightarrow{ClCO_2C_2H_5} C_6H_5CH_2O\overset{O}{\overset{\|}{C}}-NHCH\overset{O}{\overset{\|}{C}}-NHCH\overset{O}{\overset{\|}{C}}\overset{O}{\overset{\|}{O}}C_2H_5 \xrightarrow{H_2NCH_2CO_2H}$

                              $\overset{|}{CH_3} \qquad \overset{|}{CH_3}$

$C_6H_5CH_2O\overset{O}{\overset{\|}{C}}-NHCH\overset{O}{\overset{\|}{C}}-NHCH\overset{O}{\overset{\|}{C}}-NHCH_2CO_2H \xrightarrow{H_2, \ Pd}$

                        $\overset{|}{CH_3} \qquad \overset{|}{CH_3}$

$H_2NCH\overset{O}{\overset{\|}{C}}-NHCH\overset{O}{\overset{\|}{C}}-NHCH_2CO_2H$

        $\overset{|}{CH_3} \qquad \overset{|}{CH_3}$

18.34   (a), (d), (e)

18.35   The protein undergoes denaturation. The hydrogen bonding is disrupted, the globules unfold, and the protein material precipitates.

18.36   (a) The acetic acid in vinegar forms hydrogen bonds with protein groups, disrupting the hydrogen bonds that hold the protein molecules together.

   (b) Aqueous sucrose can disrupt the hydrogen bonding and tenderize the meat to some extent, but not to the degree that the strongly hydrogen bonding acetic acid can.

18.37   (a) 4   (b) 3, 5, 6   (c) 6   (d) 2   (e) 1   (f) 7

18.38   (b)

18.39 $HO_2CCH_2CH_2CHCO_2H$ + $H_3\overset{+}{N}CHCO_2H$ + $H_2\overset{+}{N}$—$CHCO_2H$ + NH_4^+ Cl^-

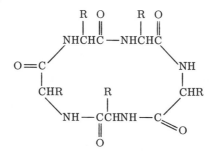

18.40   The peptide is cyclic. For example:

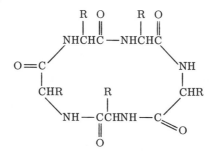

18.41   Lysine has an $\underline{sp}^3$ N on its side chain, while histidine has two ring nitrogens. The NH ring nitrogen is not basic (cf. pyrrole, page 762 of text). The other nitrogen, although basic, is in the $\underline{sp}^2$ state and therefore is less basic than the N of lysine.

18.42   (a) Because the reaction is slow for acidic and basic D-amino acids, but fast for neutral D-amino acids, we conclude that electronic effects are more important in determining the rate.

   (b) For the same reason, we predict that the binding site is nonpolar.

   (c) D-Cysteine, a neutral amino acid, undergoes oxidation faster than D-arginine, a basic amino acid.

18.43   The side chain of arginine is more basic than that of lysine because of resonance-stabilization of the cation.

lysine: $RCH_2NH_2 + H_2O \rightleftharpoons RCH_2\overset{+}{N}H_3 + OH^-$

arginine: $\underset{\overset{\|}{RNHCNH_2}}{\overset{NH}{}} + H_2O \rightleftharpoons OH^- + \underset{\overset{\|}{RNHCNH_2}}{\overset{\overset{+}{N}H_2}{}} \longleftrightarrow RNH\overset{\overset{NH_2}{|}}{C}\!=\!\overset{+}{N}H_2 \longleftrightarrow$

$$RNH\!=\!\overset{\overset{NH_2}{|}}{\underset{}{C}}NH_2$$

18.44   pro-lys-gly $\underset{cyS\text{-}pro}{\overset{cyS}{|}}$ $\underset{cyS}{\overset{tyr\text{-}cyS}{|}}$

tyr-phe-glu   phe-glu-asp $\underset{asp\text{-}cyS}{\overset{cyS}{|}}$

18.45 $H_2NCH \overset{CO_2H}{\underset{CH_2CH_2CH_2NH \diagup CNH_2}{|}}$ $\underset{\text{enzyme}}{\overset{H_2O}{\longrightarrow}}$ $H_2\overset{O}{\overset{\|}{N}}CNH_2 + H_2NCH\overset{CO_2H}{\underset{CH_2CH_2CH_2NH_2}{|}}$

ornithine

18.46 $\overset{\sim\!\!\sim}{\underset{CH_2CH_2CH_2\underset{..}{N}H_2}{|}} + \overset{\delta- \quad \delta+ \quad \delta-}{H_2N\!-\!C\!\equiv\!\underset{..}{N}:} \longrightarrow \overset{\sim\!\!\sim}{\underset{CH_2CH_2CH_2\overset{+}{N}H_2}{|}} \longrightarrow$

$H_2NC\!=\!\underset{..}{N}:^-$

$\overset{\sim\!\!\sim}{\underset{CH_2CH_2CH_2NH}{|}}$

$H_2N\overset{|}{C}\!=\!NH$

18.47   glu-cyS-gly $HO_2CCHCH_2CH_2\overset{O}{\overset{\|}{C}}\!-\!NHCH\overset{O}{\overset{\|}{C}}\!-\!NHCH_2CO_2H$

$\underset{NH_2}{|}$ $\underset{CH_2SH}{|}$

Chapter 19

Lipids and Related Natural Products

Some Important Features

Some important lipids are the edible fats and oils, terpenes, and steroids. Animal fats are triglycerides of fatty acids containing few carbon-carbon double bonds, while vegetable oils are triglycerides of fatty acids containing a greater number of carbon-carbon double bonds. Saponification of fats or oils yields soaps, RCO_2Na, where R is a long alkyl or alkenyl chain. Twenty-carbon fatty acids are used to biosynthesize prostaglandins (Section 19.5).

Phospholipids are compounds with hydrocarbon chains and a dipolar phosphate-amine group. These compounds form part of cell walls.

Terpenes are compounds with structures of head-to-tail isoprene units.

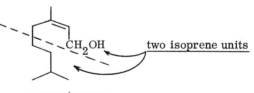

two isoprene units

a monoterpene

Terpenes are found in relatively large amounts in plants. In animals, they are intermediates in the biosynthesis of steroids. Some terpenes act as pheromones (Section 19.7).

Steroids are compounds containing the following ring system:

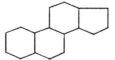

The stereochemistry of steroids is discussed in Section 19.8. Most steroids have trans-ring junctures; the bile acids are a notable exception.

Some steroids act as hormones. A few hormonal steroids are discussed in the text, along with cholesterol and vitamin D.

Reminders

Although many lipids have complex structures, keep in mind that the reactions of lipids are predictable from their functional groups. For example, triglycerides are esters and may also contain carbon-carbon double bonds. Their reactions are typical of these functional groups. Sphingomyelin (Section 19.4) contains a carbon-carbon double bond, a hydroxyl group, an amide group, an inorganic ester group, and a quaternary nitrogen. Its reactions are typical of these functional groups.

Answers to Problems

19.10

$$\begin{array}{l} CH_2O_2C(CH_2)_7CH{=}CHCH_2CH{=}CH(CH_2)_4CH_3 \\ CHO_2C(CH_2)_7CH{=}CHCH_2CH{=}CH(CH_2)_4CH_3 \\ CH_2O_2C(CH_2)_7CH{=}CHCH_2CH{=}CH(CH_2)_4CH_3 \end{array} \xrightarrow[\text{Ni}]{6\ H_2} \begin{array}{l} CH_2O_2C(CH_2)_{16}CH_3 \\ CHO_2C(CH_2)_{16}CH_3 \\ CH_2O_2C(CH_2)_{16}CH_3 \end{array}$$

19.11 (a)
$$\overset{\overset{\textstyle OH}{|}}{HOCH_2CHCH_2OH} + 2\ CH_3(CH_2)_{16}CO_2^-\ Na^+ + CH_3(CH_2)_5CH{=}CH(CH_2)_7CO_2^-\ Na^+$$

(b)
$$\overset{\overset{\textstyle OH}{|}}{HOCH_2CHCH_2OH} + 2\ CH_3(CH_2)_{16}CH_2OH + CH_3(CH_2)_{14}CH_2OH$$

(c)
$$\begin{array}{l} CH_2O_2C(CH_2)_7CHBrCHBr(CH_2)_5CH_3 \\ CHO_2C(CH_2)_{16}CH_3 \\ CH_2O_2C(CH_2)_{16}CH_3 \end{array} \quad \text{or} \quad \begin{array}{l} CH_2O_2C(CH_2)_{16}CH_3 \\ CHO_2C(CH_2)_7CHBrCHBr(CH_2)_5CH_3 \\ CH_2O_2C(CH_2)_{16}CH_3 \end{array}$$

19.12 (a) no reaction   (b) $CH_3(CH_2)_5CO_2H + HO_2C(CH_2)_7CO_2H$

(c) $CH_3(CH_2)_4CO_2H + CH_2(CO_2H)_2 + HO_2C(CH_2)_7CO_2H$

(d) $CH_3CH_2CO_2H + CH_2(CO_2H)_2 + HO_2C(CH_2)_7CO_2H$

19.13 (a) Tripalmitolein decolorizes Br_2/CCl_4 solution; tripalmitin does not.

(b) Beeswax is hydrolyzed to a fatty acid and a water-insoluble monoalcohol; beef fat is hydrolyzed to a fatty acid and glycerol (water-soluble). Also, beef fat will consume approximately twice as much NaOH per gram when saponified.

(c) Paraffin wax (a mixture of alkanes) does not undergo hydrolysis.

(d) Linoleic acid neutralizes dilute aqueous NaOH; linseed oil does not.

(e) Sodium palmitate precipitates with Ca^{2+}; sodium p-decylbenzenesulfonate does not.

(f) A vegetable oil is hydrolyzed to a fatty acid and glycerol; a motor oil (a mixture of hydrocarbons) does not undergo hydrolysis.

19.14 $CH_3(CH_2)_{12}CO_2H$

$$CH_2O_2C(CH_2)_{12}CH_3$$
$$CHO_2C(CH_2)_{12}CH_3$$
$$CH_2O_2C(CH_2)_{12}CH_3$$

19.15 $C_{15}H_{31}CO_2H + C_{15}H_{31}CH_2OH \xrightarrow[\text{heat}]{H^+} C_{15}H_{31}CO_2C_{16}H_{33} + H_2O$

19.16   (a), (b), (c), (e). In each case, the molecule contains an ionic end as well as a hydrocarbon end.

19.17
$$CH_2O_2C(CH_2)_{14}CH_3$$
$$CHO_2C(CH_2)_{14}CH_3$$
$$CH_2O_2C(CH_2)_7CH=CH(CH_2)_7CH_3 \leftarrow$$

Oleic acid must be attached to carbon 1 of glycerol; otherwise, carbon 2 would not be chiral.

19.18   tristearin

$$\xrightarrow[\text{(2) } H^+]{\text{(1) NaOH, } H_2O, \text{ heat}} CH_3(CH_2)_{16}CO_2H$$

$$\xrightarrow[\text{heat pressure}]{H_2, \text{ catalyst}} CH_3(CH_2)_{17}OH$$

$$\left.\vphantom{\begin{array}{c}1\\2\end{array}}\right\} \xrightarrow[\text{heat}]{H^+} CH_3(CH_2)_{16}CO_2(CH_2)_{17}CH_3$$

19.19   (a) $CH_3(CH_2)_7CH=CH(CH_2)_7CO_2Na + CH_3(CH_2)_{16}CO_2Na +$

$$Na_3PO_4 + HOCH_2CH_2\overset{+}{N}(CH_3)_3 \ Cl^- + HOCH_2\overset{\overset{\displaystyle OH}{|}}{C}HCH_2OH$$

(b) $CH_3(CH_2)_7CH=CH(CH_2)_7CO_2Na + CH_3(CH_2)_{16}CO_2Na +$

$$Na_3PO_4 + HOCH_2CH_2NH_2 + HOCH_2\overset{\overset{\displaystyle OH}{|}}{C}HCH_2OH$$

19.20 $CH_3(CH_2)_{22}CO_2H +$

19.21 (a) 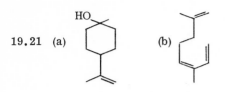 (b)

19.22 (a) and (c), sesquiterpenes   (b) monoterpene

19.23    sesquiterpene

19.24 (a) 3   (b) 1   (c) 2

19.25 (a) bicyclo[2.1.1]hexane   (b) bicyclo[4.1.0]heptane
       (c) bicyclo[2.2.2]octane

19.26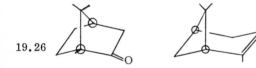

19.27 (a) 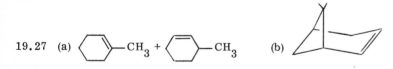 (b)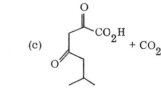

       In (b), the double bond cannot go to the bridgehead.

19.28 (a)   (b)

19.29 (a) 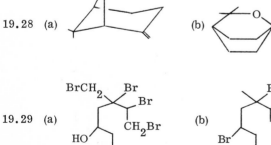 (b) (c) $+ CO_2$

19.30 (a) trans   (b) trans   (c) trans   (d) trans   (e) cis   (f) cis
       (g) trans   (h) cis

19.31 (a) (2) because all ring substituents are equatorial.

(b) (1) because the ring juncture is <u>trans</u>.

(c) (1) because the OH group is equatorial.

19.32 (a)   (b)

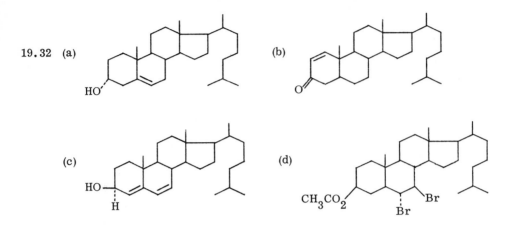

(c)   (d)

19.33 Estradiol is a phenol and can be extracted from the mixture by aqueous NaOH.

19.34

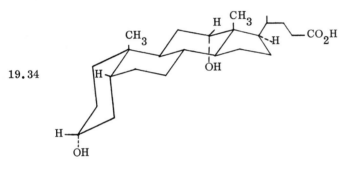

19.35 (a)

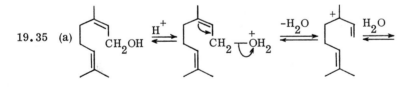

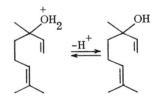

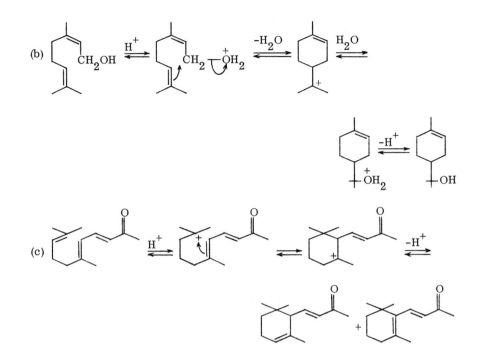

(b)

(c)

19.36 (a) none because the ^{14}C is lost as $^{14}CO_2$

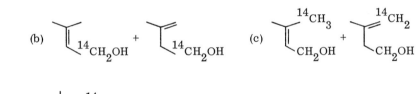

(b)

(c)

19.37

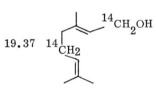

19.38

19.39

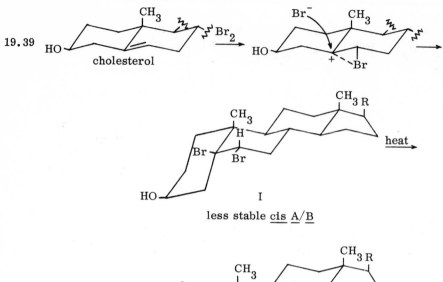

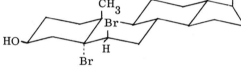

I

less stable <u>cis</u> <u>A/B</u>

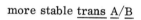

more stable <u>trans</u> <u>A/B</u>

Chapter 20

Spectroscopy II:
Ultraviolet Spectra, Color and Vision, Mass Spectra

Some Important Features

Ultraviolet and visible spectra arise from the promotion of pi electrons, especially pi electrons that are part of a conjugated system. In general, the larger the conjugated system, the longer is the wavelength of light absorbed. If the conjugated system is long enough, the compound appears colored.

$$C=C \qquad C=C-C=C \qquad C=C-C=C-C=C$$
Absorbed light of increasing λ
and decreasing energy

The quantity of radiation absorbed (the <u>molar absorptivity</u> ϵ) is calculated from the formula $\epsilon = \underline{A/c\ l}$ (Section 20.2). Transitions of the $\pi \rightarrow \pi^*$ type generally have a high value for ϵ compared to $\underline{n} \rightarrow \sigma^*$ or $\underline{n} \rightarrow \pi^*$ transitions.

Color usually arises from the absorption of specific wavelengths from the full visible range (white light). Reflection of the remaining, nonabsorbed wavelengths to the eye results in color vision.

Some colored organic compounds are discussed in Sections 20.6, 20.7, and 20.8. Of particular interest are the indicators that change color depending on the pH of the solution. An acid-base reaction of an indicator molecule results in a change in the length of the conjugated system and a change in the wavelength of absorption.

Fluorescence and chemiluminescence are mentioned briefly (Section 20.9). The biochemical transformations within the eye that lead to vision are discussed in Section 20.4A.

<u>Mass spectra</u> arise from cleavage of molecules into ions and ion-radicals when the molecules are bombarded with high-energy electrons. Loss of a single electron gives rise to the <u>molecular ion,</u> often the farthest peak to the right in a mass spectrum.

$$\overset{\displaystyle \overset{\cdot\cdot\,\cdot\cdot}{O}}{\underset{CH_3\overset{\|}{C}CH_3}{}} \xrightarrow{\ -e^-\ } \left[\overset{\displaystyle \overset{\cdot\cdot}{O}\cdot}{\underset{CH_3\overset{\|}{C}CH_3}{}} \right]^{+}$$

molecular ion
<u>m/e</u> = 58

Fission of the molecular ion often occurs at a branch, an electronegative atom, or a carbonyl group:

Small molecules, such as H_2O, can be lost:

$$[RCH_2CH_2OH]^{\overset{+}{\cdot}} \xrightarrow{-H_2O} [RCH{=}CH_2]^{\overset{+}{\cdot}}$$

An alkene molecule may be lost in a McLafferty rearrangement (Section 20.13D).

Reminders

In a problem concerning the color change of a compound with a change in pH: (1) look for the most acidic or basic group in the molecule; (2) write the acid-base reaction; (3) write resonance structures for reactant and product.

In mass spectral problems, m/e is the <u>mass of the ion or ion-radical</u> divided by the <u>ionic charge</u> (usually +1).

To determine fragmentation patterns, write the structure of the molecular ion. Mark all likely positions of cleavage, note any small molecules that could be lost, then write the structures of the fragments, keeping track of electrons.

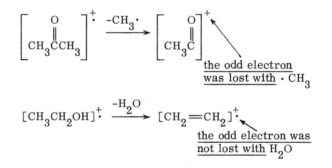

Answers to Problems

20.17   In the uv spectra, the tetraene would show a λ_{max} at a longer wavelength than the triene.

20.18   (a), because the OH electrons are in conjugation with the other pi electrons in one case.

   (d), because the carbonyl group in conjugation with the double bond exhibits different absorption from that of the conjugated diene.

(e), because one structure contains conjugation and the other does not.

(f), because one structure is a conjugated triene, while the other is a conjugated diene.

In (b) and (c), the pi systems are the same and absorb radiation outside the usual uv range.

20.19 $\epsilon = \dfrac{A}{c\ l} = \dfrac{7.4}{(0.5)(1.0)} = 15$

20.20 $\epsilon = \dfrac{0.75}{(0.038)(1.0)} = 20$

20.21 (a) $n \to \pi^*$ and $\pi \to \pi^*$ transitions   (b) the transition at 219 nm

20.22 (a) $\pi \to \pi^*$ (b) $\pi \to \pi^*$ and $n \to \pi^*$ (c) $\pi \to \pi^*$ and $n \to \pi^*$
(d) $\pi \to \pi^*$ and $n \to \pi^*$.

(All could exhibit $\sigma \to \pi^*$ transitions also.)

20.23 (a), (c), (b), (d), the same order as that of increasing conjugation.

20.24 A conjugated diene, $CH_2{=}CHCH{=}CHCH_3$ or $CH_2{=}\overset{\overset{\displaystyle CH_3}{|}}{C}CH{=}CH_2$

20.25 Because of van der Waals repulsions, <u>cis</u>-stilbene cannot be planar; therefore, the degree of conjugation is less and the wavelength of uv absorption is less.

20.26

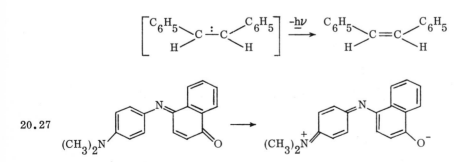

rotation around sigma bond

20.27

20.28

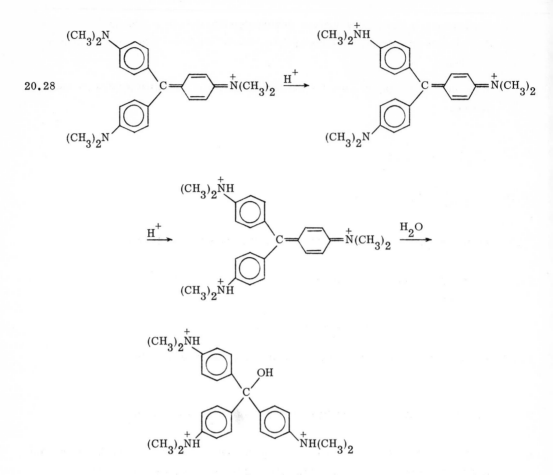

20.29 At pH 6, the structure is as shown in the problem. At pH 9:

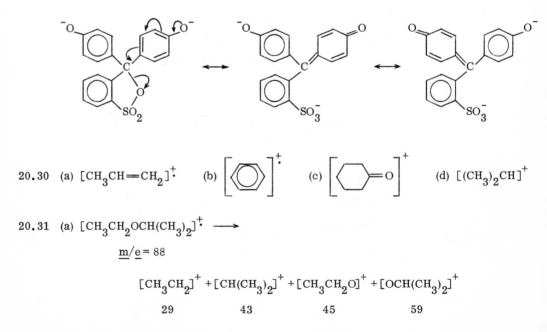

20.30 (a) $[CH_3CH=CH_2]^{+\cdot}$ (b) $[\bigcirc]^{+\cdot}$ (c) $[\bigcirc=O]^{+}$ (d) $[(CH_3)_2CH]^{+}$

20.31 (a) $[CH_3CH_2OCH(CH_3)_2]^{+\cdot} \longrightarrow$

$\underline{m/e} = 88$

$[CH_3CH_2]^{+} + [CH(CH_3)_2]^{+} + [CH_3CH_2O]^{+} + [OCH(CH_3)_2]^{+}$

       29   43   45   59

(b) $[CH_3CH_2OCH_2CH(CH_3)_2]^{\ddagger} \longrightarrow [CH_3CH_2]^+ + [CH_2CH(CH_3)_2]^+$

$\underline{m/e} = 102$ 29   57

$+ [CH_3CH_2OCH_2]^+ + [CH(CH_3)_2]^+ + [CH_3CH_2O]^+ + [OCH_2CH(CH_3)_2]^+$

59   43   45   73

(c) $[(CH_3)_2CHCl]^{\ddagger} \longrightarrow [(CH_3)_2CH]^+$

$M^{\ddagger} = 78$ 43

$M + 2 = 80$

(d) $[(CH_3)_2CHCH_2CH_2CH(CH_3)_2]^{\ddagger} \longrightarrow [(CH_3)_2CH]^+ + [CH_2CH_2CH(CH_3)_2]^+$

$\underline{m/e} = 114$ 43   71

(e) $[(CH_3)_2CHOH]^{\ddagger} \longrightarrow [CH_2\!=\!CHCH_3]^{\ddagger} + [(CH_3)_2CH]^+$

$\underline{m/e} = 60$ 42   43

(f) $\left[\bigtriangleup\!\!-CH_2CH_2CH_2CHO\right]^{\ddagger} \longrightarrow [CH_2\!=\!CHOH]^{\ddagger}$

$\underline{m/e} = 140$ 44

20.32 (a) $[CH_3CH_2CH_2CH_3]^{\ddagger} \longrightarrow [CH_3CH_2CH_2CH_2]^+ \longrightarrow$

58   57

$[CH_3CH_2CH_2]^+ \longrightarrow [CH_3CH_2]^+ \longrightarrow [CH_3]^+$

43   29   15

(b) $\left[C_6H_5\overset{\displaystyle O}{\overset{\|}{C}}NH_2\right]^{\ddagger} \longrightarrow \left[C_6H_5\overset{\displaystyle O}{\overset{\|}{C}}\right]^+ \longrightarrow [C_6H_5]^+$

121   105   77

(c) $[CH_3CH_2CH_2Br]^{\ddagger} \longrightarrow [CH_3CH_2CH_2]^+ \longrightarrow [CH_3CH_2]^+ \longrightarrow [CH_3]^+$

$M^{\ddagger} = 122$ 43   29   15

$M + 2 = 124$

20.33   2 HO— 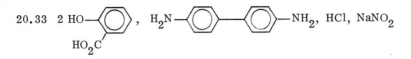 , H_2N— —NH$_2$, HCl, NaNO$_2$

Make the diazonium salt of the biphenyl compound and couple it with salicylic acid.

20.34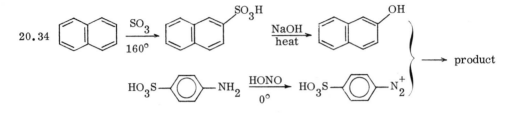

⟶ product

20.35   The addition of each alkyl substituent to the conjugated system adds about 5 nm to the λ_{max}.

20.36   (a) 232 nm   (b) 232 nm   (c) 262 nm

20.37   (a) 229 nm   (b) 234 nm   (c) 249 nm

20.38   (b), (c), (d). All are methyl ketones and yield acetyl fragments (43), $[C_4H_9]^+$ (57), butanoyl fragments (85), and a molecular ion (100).

20.39   CH$_3$CCH$_3$ (with O double bonded to the central carbon)

20.40   (a) acetone
(b) The λ_{max} at 230 nm is from a $\pi \rightarrow \pi^*$ transition; that at 320 nm is from a $\underline{n} \rightarrow \pi^*$ transition.
(c) The $\underline{n} \rightarrow \pi^*$ transition is observed; the $\pi \rightarrow \pi^*$ is below 200 nm and cannot be seen.

20.41   A, ClCH$_2$CH$_2$CH$_2$OH   B, CH$_3$—⟨⟩—CN

C, C_6H_5CCH$_2$CH$_2$CH$_3$ (with O double bonded to carbon)   D, C_6H_5I